Smart Universities in Smart Cities

Smart Universities in Smart Cities: Shaping the Future of Education and Urban Innovation focuses on how higher education institutions are adapting to the challenges of the digital age in a world increasingly influenced by technology and sustainability. Universities are becoming an indispensable element of smart cities, driving forward innovation, sustainability, and urban living.

The book explores how emerging technologies such as AI, IoT, and blockchain are transforming campuses into dynamic, data-driven ecosystems. Both of these dimensions are studied through data analytics, case studies, and futuristic thinking perspectives to identify opportunities and challenges of the establishment of smart universities within the broader ecosystem of smarter cities. The book offers a holistic approach to planning educational institutions, covering issues such as sustainable campus development, digital learning environments, and smart mobility solutions.

As universities undergo digital transformation, they advance not only learning but also the larger role of academia in society. The book envisions the future, where intelligent campuses act as centres of knowledge, collaboration, and sustainable development, propelling the world into a smarter and more inclusive reality for future generations. It will be of interest to researchers and students of urban planning and sustainability studies, as well as to urban planners and policymakers.

Joanna Rosak-Szyrocka is Assistant Professor, Erasmus+ coordinator at the Faculty of Management, Czestochowa University of Technology, Poland. She has been a participant in multiple Erasmus+ teacher mobility programs: Italy, UK, Slovenia, Hungary, Czech Republic, Slovakia, and France.

Radosław Wolniak is a full professor at the Department of Economics and Informatics at the Silesian University of Technology, Poland. For over 20 years, he has been dealing with the issues of management, including the Industry 4.0, Smart City, digitalization, and artificial intelligence. His achievements include over 600 publications issued at national and international publishers, as well as many undertakings and projects implemented in cooperation with the business environment.

Routledge Research in Planning and Urban Design

Routledge Research in Planning and Urban Design is a series of academic monographs for scholars working in these disciplines and the overlaps between them. Building on Routledge's history of academic rigour and cutting-edge research, the series contributes to the rapidly expanding literature in all areas of planning and urban design.

The Protection of Green Spaces for Climate Change Adaptation
Planning Systems, Policies and Instruments
Edited by Maciej J. Nowak

Exclusion in Smart Cities
Assessment and Strategy for Strengthening Resilience
Izabela Jonek-Kowalska and Radosław Wolniak

The Green Belt, Housing Crises and Planning Systems
Charles Edward Goode

Displaced Urbanism
Edited by Gihan Karunaratne

Resilient Urbanism
Edited by Gihan Karunaratne

Contracting with Citizens
How Morals, Politics, and Laws Shape Cities
E. W. (Michiel) Stapper

Smart Universities in Smart Cities
Shaping the Future of Education and Urban Innovation
Joanna Rosak-Szyrocka and Radosław Wolniak

For more information about this series, please visit: www.routledge.com/Routledge-Research-in-Planning-and-Urban-Design/book-series/RRPUD

Smart Universities in Smart Cities

Shaping the Future of Education and Urban Innovation

Joanna Rosak-Szyrocka and Radosław Wolniak

Routledge
Taylor & Francis Group
LONDON AND NEW YORK

Designed cover image: Getty Images

First published 2025
by Routledge
4 Park Square, Milton Park, Abingdon, Oxon OX14 4RN

and by Routledge
605 Third Avenue, New York, NY 10158

Routledge is an imprint of the Taylor & Francis Group, an informa business

British Library Cataloguing-in-Publication Data
A catalogue record for this book is available from the British Library

ISBN: 978-1-041-03011-9 (hbk)
ISBN: 978-1-041-03014-0 (pbk)
ISBN: 978-1-003-62182-9 (ebk)

DOI: 10.4324/9781003621829

Typeset in Times New Roman
by SPi Technologies India Pvt Ltd (Straive)

Contents

Figures

Tables

Introduction

One of the most important aspects of technology in education is its ability to level the field of opportunity for students.
John King, former U.S. Secretary of Education

Tell me and I forget, show me and I may remember, let me take part and I understand
Confucius

We would like to start our introduction from these two quotes that are immensely rich in meaning for what this book attempts to explore in all its dimensions – the power of integrating technology in education; breaking down physical barriers and opening up the doors to new learning outcomes. The authors of the book *Smart Universities in Smart Cities: Shaping the Future of Education and Urban Innovation* observe the democratizing effect of technology in education and emphasize that by using the internet and local devices to provide learning resources and collaboration platforms and practicing formative assessments, we can create classroom ecosystems that provide equal access to knowledge and opportunities to every learner. Smart universities are creating equal opportunity through technological advancements such as AI, IoT, and cloud computing, which are facilitating personalized learning pathways for all students, along with tailored resources and real-time support.

The digital campus initiative with the intent to potentially create practical classroom scenarios for students is not a smart university, it should be the learning environment based on student engagement, students must experience the environment where they can interact with the tools in a such way that enables them to learn better, faster, and more effectively. This interactive, participatory method mirrors the tenets of smart education in which learning is now no longer passive but experiential and dynamic. The interaction between these two poles can best describe the heart of this book: that the essence of university smart technologies is not merely technical innovation but a catalyst for learning equity and engaging, participative learning.

DOI: 10.4324/9781003621829-1

This sets the stage to pave the way for a discussion on how technology in education can, when thoughtfully implemented, enrich academic experiences while also nurturing scenes that are inclusive, participatory, and transformative building blocks for students everywhere. The influx of other smart technologies on campuses enables us to offer all students innovative paths to learn, create, and excel in ways that once seemed impossible, inherent in a combination of time-tested truths and great educational innovation in the present day. However, with smart technology making its way onto university campuses, it creates a unique opportunity to assist in closing education gaps through personalized learning pathways, real-time support, and tailored resources. And it gives a whole new way of looking at what being 'smart' means in education:

S – sustainable development,
M – modern,
A – advanced technology,
R – resilient,
T – transformative.

Sustainable Development: This is based on international directives for environmental, social, and economic sustainability, which are fundamental to every smart system, specifically to smart cities and universities. This means that the systems are created with ecological and social benefits over time.

Modern: This focuses on cutting-edge approaches and practices that include the most relevant technologies and methodologies to current educational and urban challenges. It leads to contemporary infrastructure, policies, and educational systems that are suited for today's digital and societal climate.

Advanced Technology: This defines the most advanced tools, including AI, IoT, and blockchain, which are crucial for developing smart campuses and cities. By leveraging big data, automation, and decision-making as services, these technologies foster efficiency and drive innovation.

Resilient: Smart systems should be flexible to unknown challenges and able to recover quickly from failures. This quality highlights the need for resilient designs despite environmental, technological, and social disruptions as we expect that smart campuses and cities can flourish even in complex environments.

Transformative: Smart environments, whether in universities or cities, should be transformative, changing the way people learn, the management of resources, or how urban living is organized. It emphasizes the visionary imagination behind 'smart', seeking new systems that not only enhance but also transform education and cities.

This book links the ramifications of technology for democratizing education to insights about the lessons of our teaching and learning – making way for the larger theme of how universities are remaking themselves in the face of change introduced by both technology and the society in which they exist.

This book is about the way technology and education are finding themselves together at the same intersection in an era when universities are busy trying to redefine themselves in the face of digital transformation. Exploring the potential of smart campuses, we seek to identify how this evolving integration can lead to a richer, more inclusive, and dynamic learning experience. The future of education isn't simply using technology, it's a complete rethinking of how we learn, grow, and teach inside a globally linked world.

Because the authors wrote the book with such passion, we hope the reader will find it interesting.

Dr Joanna Rosak-Szyrocka
Full Prof. Radosław Wolniak

1 Overview of Smart Universities and Smart Cities

1.1 Evolution of Smart Concepts: From Cities to Campuses

According to UNEP (2019), 66% of people on Earth are expected to live in cities by 2050, up from around 54% at the present time. This suggests that the number of urban dwellers worldwide may increase by 2.4 billion. As a consequence of this, current urban settings will unavoidably need to be significantly expanded, and new ones will need to be created. Although cities only make up 2% of the planet's area, they utilize over 75% of its natural resources. According to Manager (2018), the amount of material used by cities is expected to increase from 40 billion tons in 2010 to around 90 billion tons by 2050. Water, food, fossil fuels, raw materials, and basic energy are a few of these resources ('Sustainable, Resource Efficient Cities - Making It Happen | Department of Economic and Social Affairs', 2024). Cities are thus anticipated to have difficulties with regard to development, performance, competitiveness, and the standard of living of their citizens ('Smart cities: Digital solutions for a more livable future', 2018). According to Washburn et al. (2010), growing urbanization raises a number of issues, including deteriorating liveability due to issues with waste management, limited resources, air pollution and traffic congestion that pose health risks to people, and aging public infrastructure. Among the potential answers to these problems is the idea of a 'smart city'.

Some definitions define 'smart' as being generally extremely excellent at learning new things, exhibiting intelligent judgement, and responding quickly to difficulties. A system may be deemed 'smart' when it is able to independently provide services in accordance with the changing demands of its users (Kwok, 2015). In order to manage metropolitan areas and their resources and assets efficiently, the term 'Smart-City' was coined to describe the use of autonomous environmental data gathering and processing (Schaffers et al., 2011). The widespread use of information and communication technologies (ICTs) and the Internet of Things (IoT) paradigm (Zanella, Bui, Castellani, Vangelista, & Zorzi, 2014), in which a large number of dispersed objects are linked to exchange their collected data, assists this strategy. Since telecommunication technologies are as important as the associated sensing and processing operations, the Smart-City's need for connection highlights their significance (Pan et al., 2023). Numerous cities worldwide have adopted the Smart-City idea in an effort to enhance the well-being and

DOI: 10.4324/9781003621829-2

quality of life of their residents, as well as to lower emissions, improve governance, and boost energy efficiency (Gascó-Hernandez, 2018; Ojo, Curry, & Zeleti, 2015; Sotres, Santana, Sanchez, Lanza, & Munoz, 2017). The notion of a smart city is founded on the idea of smart infrastructure, which is the connectivity of digital components to produce and utilize data intelligently to enhance system performance via resource efficiency (Omotayo, Awuzie, Ajayi, Moghayedi, & Oyeyipo, 2021). Omotayo, Olanipekun, Obi, and Boateng (2020); Omotayo, Moghayedi, Awuzie, & Ajayi, 2021) have identified the essential infrastructure components required for smart city solutions, which include but are not limited to a sustainable environment, complete IT digitalization, effective waste management, energy management, efficient urban mobility, e-governance, telemedicine/education, and a steady supply of electricity. The environmental, economic, and sociological aspects of sustainability were all included in many definitions, while others only looked at one or two of them. There doesn't seem to be a single, universal definition of what a 'smart city' is (O'Grady & O'Hare, 2012), suggesting that definitions of smart cities are diverse. Table 1.1 presents the definition of a smart city.

Table 1.1 Chosen definition of a smart city

Author/s	*Definition*
('New Report: Building Energy-Smart Cities is Fastest, Easiest Way to Address Indian Energy Crisis', 2012)	A smart city is a city that may aim to make itself 'smarter', more sustainable, efficient, equitable, and liveable
('Smart Cities Study: International study on the situation of ICT, innovation and Knowledge in cities', 2012)	Smart city is a city that utilizes ICT to create more interactive and efficient components and utilities of critical infrastructure
(Falconer & Mitchell, 2012)	Smart cities adopt scalable solutions that utilize ICT to boost efficiency, decrease costs, and improve quality of life
(Alizadeh, 2017; WebWire, 2024)	A city can become smart by optimally using all the available interconnected information to comprehend and regulate its operations and optimize the utilization of available resources
(Zygiaris, 2013)	Smart city is a certain intellectual ability that addresses several innovative socio-technical and socio-economic aspects of growth
('The smart city vision: how innovation and ICT can build smart, "livable", sustainable cities', 2010)	Smart cities strive to improve city services and urban management for the citizens by creating a socially advanced environment. The ultimate goal of these processes is to improve the sustainability and liveability of the city
(Hollands, 2020)	Smart City is a city that aims at connecting the physical, IT, social, and business infrastructures to leverage the collective intelligence of the city

Source: own study.

The definitions therefore address several aspects of smart city development, from the use of ICT, user communication, e-governance, and equitable development to sustainability and education. Besides differences in substance, there is a disparity in how sustainability is prioritized as a smart city objective across definitions that focus on sustainability. Sustainability and liveability are considered to be among the main objectives of smart cities in many of the definitions that are put out. In almost one-third of the definitions given, sustainability is included as one of the secondary objectives of smart cities, along with governance, liveability, and resource efficiency. Sustainability was not emphasized as a tertiary aim in many definitions, which lessened its significance in the agenda for smart cities. In tertiary definitions, the main objective is governance and quality of life.

Due to the influx of digital native students who want a more seamless experience that fosters success, higher education is undergoing complex upheaval (Rico-Bautista et al., 2021; Rosak-Szyrocka, Zywiolek, Zaborski, Chowdhury, & Hu, 2022; Villegas-Ch, Palacios-Pacheco, & Luján-Mora, 2019). People's life might be changed in this situation by a smart university (Halawani & Mohandes, 2003; Rosak-Szyrocka & Tiwari, 2023). In fact, an increasing number of academics are looking to university campuses as excellent locations to test new services and methods for smart cities, creating what is known as a 'smart campus' (Verstaevel, Boes, & Gleizes, 2017).

The Smart-Campus that results from implementing the Smart-City concept in a university region may expand with community participation and provide significant benefits. The administration, governance, sustainability, and educational activities of campuses, which are their own 'small cities', may be greatly improved in terms of energy and water efficiency, emissions, mobility, health and well-being, nature, and instruction (Fortes et al., 2019; Omotayo, Awuzie, et al., 2021. When it comes to incorporating the Smart-Campus idea into educational activities, several research studies concentrate on implementing advanced smart capabilities that have the potential to enhance education, like an IoT-based Smart-Campus concept with an emphasis on smart parking, smart rooms, and smart teaching (Sari, Ciptadi, & Hardyanto, 2017). Nonetheless, remote learning (e.g., via video streaming) is the primary goal of the planned learning-related applications. A number of options are identified by Majeed et al. (Majeed & Ali, 2018) for implementing the smart idea on campuses, with a focus on parking, security, classroom assistance, and education. Other studies focus on the intelligent analysis and/or support of educational activities, including the use of game-based learning (Zhai, Dong, & Yuan, 2018), multimedia conference assistance (Zhang, Zhang, & Shi, 2018), student health (Gao et al., 2018), and smartphone apps (Lin, Chen, Shieh, Lin, & Yen, 2018). According to Alvarez-Campana et al. (Alvarez-Campana, López, Vázquez, Villagrá, & Berrocal, 2017), an IoT platform for monitoring human and environmental fluxes is presented on university campuses.

The goal of a Smart Campus, according to Pinto et al. (Pinto, Romano, & Tomoto, 2017), is to offer a cutting-edge, multidisciplinary educational experience that uses the campus as a prototyping space and supports the education of citizen engineers by solving real-world problems. This will enable them to multiply in an urban setting driven by innovative thinking, with a dedication to the environment and social responsibility, and be replicable in other nations. Using and manipulating a lot of data remotely, smart campus projects are centred on the idea of the IoT and how it interacts with educational buildings via smart network grids (Omotayo, Moghayedi, et al., 2021). The project's foundation is the development of smart buildings or the conversion of existing campus structures to provide a sustainable setting necessary for learning (Chen, Jiang, Yang, Yang, & Man, 2017). According to Omotayo et al. (Omotayo et al., 2020; Omotayo, Moghayedi, et al., 2021), 5G and 6G flagship, smart grid, artificial intelligence (AI), fog computing, green computing, repurposing, energy management, and smart teaching and learning are the main forces behind the IoT–smart building interaction required for the smart campus. IoT and smart classroom technologies provide for the efficient and effective provision of citizen and e-governance services to the public, as well as social benefits from resource management within the university ecosystem (Zhang, Dong, Yip, & Swift, 2020). The usage of big educational buildings and other facilities, according to Moura et al. (Moura, Moreno, López López, & Alvarez-Campana, 2021), supports high energy consumption costs on campuses, necessitating the adoption of smart campus solutions for effective energy management systems (EMS). However, the integration of IoT offers platforms for monitoring and managing power supply in order to assess generation and demand, achieve sustainable energy usage, and deploy smart devices to regulate loads using data. For EMS to operate efficiently inside a smart campus network grid, cloud computing in conjunction with machine learning analysers is a crucial element (Omotayo, Awuzie, et al., 2021). Net-zero energy, photovoltaic, laser scanning, renewable energy, and building information modelling to building energy models are the main motivators.

As a key component of the smart city framework, smart campuses provide residents with a smart learning environment that helps them become smart workers (Kwok, 2015; Liu, Huang, & Wosinski, 2017). The growth and acceptance of smart campuses also contribute to the knowledge economy. At a compound annual growth rate (CAGR) of 19.6%, the worldwide smart education and learning market is expected to grow from its 2019 valuation of USD 191.32 billion to USD 783.48 billion by 2027 ('Smart Education and Learning Market | Global Report [2027]', 2024). Several scholars, who have created smart campuses issue, describe the idea using different methodologies. These strategies fall into three primary categories: *organizations or business process development, acceptance of smart city concepts, and technology-driven.*

As digital campuses have grown, a smart campus is an evolutionary tendency from a technology standpoint (Chuling, Xie, & Peng, 2009; Nie, 2013). To connect previously disparate systems (Nie, 2013), this development makes use of a wide range of internet-based technologies and services (Coccoli, Guercio, Maresca, & Stanganelli, 2014), such as cloud computing and IoT service providers (Manqele, Dlodlo, Coetzee, Williams, & Sibiya, 2015).

Through the integration of sensors (Cata, 2015) and sophisticated intelligence, IoT services turn commonplace items in academic settings into intelligent objects, enabling intelligent decision-making across the campus ecosystem (Owoc & Marciniak, 2013). Additionally, smart campuses allow the creation of pertinent apps and services by establishing a centralized digital nervous system (DNS), which propels a comprehensive learning cycle across campus ecosystems (Tsai & Yeh, 2015). In addition to improving graduate quality and campus performance, these technologies also make it easier for all stakeholders to actively participate in an environment that is adaptable (Coccoli et al., 2014; Rosak-Szyrocka, Apostu, & Pupp, 2024). The three main facets of this flexibility are service, management, and instruction (Chuling et al., 2009). Table 1.2 presents definitions of a smart campus.

According to the adoption strategy for smart cities, a smart campus and a smart city are comparable in that they both solve issues that both metropolitan settings and universities often encounter. Table 1.3 presents advantages between smart cities and smart universities.

As part of the same concept, a smart campus makes use of contemporary technology to assist a range of users, such as staff, students, and visitors, in carrying out different duties across the facility (Mattoni et al., 2016). The smart campus essentially operates as a small, independent city, including components such as a range of activities, user interactions, and integrated technologies (Pagliaro et al., 2016). With an emphasis on the methodical creation of

Table 1.2 Chosen definition of a smart campus

Author/s	*Definition*
Bandara (Manufacturing & Industrial Engineering Symposium (MIES)', 2016)	An initiative to use ICT within a university campus to improve the quality and performance of the services, to reduce costs and resource consumption, and to engage more effectively and actively with its members
Paola (IEEE 5th International Forum on Research and Technology for Society and Industry (RTSI)', 2019)	A digitally augmented campus where pervasive instrumented objects and spaces are made responsive to the state of the environment and its inhabitants
Muhamad (International Conference on Information Technology Systems and Innovation (ICITSI)', 2017)	The main role of smart campus is to present the dynamic services according to the needs of the users using some intelligence system
Zhang et al. (2020)	An educational environment that is penetrated with enabling technologies for smart services to enhance educational performance while meeting stakeholders' interests, with broad interactions with other interdisciplinary domains in the smart city context

Source: own study.

Table 1.3 Advantages between smart cities and smart universities

Aspect	*Advantages in smart cities*	*Advantages in smart universities*
Infrastructure integration	Efficient urban living depends on the interconnectivity of urban systems for traffic, energy, and waste management.	Big data analytics used to run efficient processes through IoT tools for maintenance, security, and campus-wide digital offerings.
Sustainability	Smart cities adopt environmentally friendly technologies and policies to promote emission control and resource conservation.	Green campuses employ renewable energy, sustainable building materials and recycling initiatives.
Technology adoption	Smart cities use IoT, AI, and big data to enhance public services such as healthcare, transportation, and utilities.	Smart Classrooms — Use of AI– will make the education experience more rewarding.
Community engagement	Digital platforms and public consultations allow residents to participate actively in governance and urban planning.	Councils and digital forums for policymaking empower students to lead, and feel ownership in the school's successful operations.
Personalization	In fields where personalization is the key that unlocks the door to higher efficiency, like real-time traffic management systems or healthcare plans customized by people's data analytics.	Customized learning pathways through adaptive learning tools and AI-based recommendations.
Knowledge sharing	Enable partnership between governments, companies, and academia to promote collaboration across different fields.	Promotes research partnerships between universities and industries to combat global education issues.
Innovation ecosystem	Start-up incubators and accelerators in national urban challenge-solving and local entrepreneurship	Can also support the creation of innovation hubs where students and faculty work together on projects to address real-world educational needs.
Data security	Focus on data protection using strong frameworks to protect the data of citizens.	Smart universities put in place secure learning management systems to protect students' data and intellectual property.
Testing and prototyping	Urban environments have been tested as platforms for smart technologies such as autonomous vehicles, smart grids, and public IoT devices.	Educational tools and smart technology solutions can be piloted on campuses as their managed environments act as a controlled setting.
Global competitiveness	Improves the appeal of a city for foreign investment, tourism, and talent retention via modern infrastructure.	Since the leading-edge technology attracts the best students and faculty, it creates a reputation for high education quality.

Sources: own study based on (El Khatib, Alhammadi, & Almulla, 2024; Liu et al., 2017; Omotayo, Moghayedi, et al., 2021).

procedures within an academic environment, the third approach conceptualizes the smart campus as an organizational or business process model. A smart campus's main idea is to utilize resources efficiently (Alghamdi & Shetty, 2016), which is accomplished by using different forms of intelligence to provide intelligent services that are of high quality (Kwok, 2015) to the campus and surrounding populations (Malatji, 2017). Overall convenience and quality of life are enhanced while operating expenses are drastically decreased (Alghamdi & Shetty, 2016; Malatji, 2017). By using cutting-edge sensor technologies, intelligence is further used to facilitate automated reporting in all facets of campus life, such as education, social interaction for teamwork, intelligent building management, and the development of a smart environment. The fundamental concept behind a smart campus is the integration of sophisticated intelligence technologies by universities to improve performance, graduate quality, and quality of life. This is achieved via the provision of important, dynamic, and user-friendly information technology services that provide automation and real-time reporting. These services include a wider range of topics than only educational activities, including energy saving, workplace operations, social engagement, and environmental management. Following case studies provide innovative instances of this progression, demonstrating smooth transitions from smart cities to smart campuses.

Case Study 1: Masdar City and Masdar Institute of Science and Technology, United Arab Emirates (UAE) – One of the most ambitious smart city projects in the world, Masdar City in the United Arab Emirates, is intended to create a zero-carbon, zero-waste metropolitan environment powered by renewable energy. The city's areas of academia and research, the Masdar Institute of Science and Technology, was established as a direct consequence of the smart infrastructure of the city, which includes energy-efficient buildings, automated waste disposal systems, and autonomous transportation. Renewable energy by city-level renewable energy techniques like solar to track and improve energy consumption and resource management, the institution has adopted advanced user-level technologies like IoT and AI.

Case Study 2: Arizona State University (ASU), USA – Arizona State University has embraced the idea of a smart city as part of the effort to make the Phoenix metropolitan region more technologically and environmentally sophisticated. ASU has integrated smart water management technologies, solar energy, and AI-driven urban planning into its campus operations in Phoenix, a city recognized for its severe weather. The campuses of ASU have integrated smart technology to enhance campus mobility systems, control energy via solar power, and minimize water use. In an arid region, the university's smart irrigation systems, which are similar to those used by the city, aid in water conservation. AI has also been used on campus to improve student services and administrative efficiency.

Case Study 3: The University of Glasgow, Scotland – Glasgow's transition to a smart city has been substantial, with projects centred on energy saving,

IoT implementation, and data analytics. All these technologies have been adopted as smart campus by the University of Glasgow, one of the top smart campuses in Europe, under the city's smart city vision. Digital learning platforms, AI-assisted building management systems, and smart energy grids have been built on among other digital initiatives and included in the transformation of the city, which aligns with the broader digital transformation of Glasgow. Using IoT devices and real-time data, the institution optimizes energy use across all campus buildings and enhances learning environments with smart classrooms and personalized learning pathways.

Case Study 4: Seoul National University (SNU), South Korea – The world's trailblazer in building smart cities, Seoul employs the foundation of AI-enabled public services technology in addition to high-speed internet infrastructure and environmentally friendly urban design to integrate technology with urban life. SNU applied many of these inventions to constitute a cutting-edge smart campus, emphasizing the power of the city's technology. Integrating the city's digital infrastructure, SNU has enhanced its operations and student services. AI and big data are being called on to predict and meet student demand, whether it is for personalized learning experiences or more efficient campus logistics, while high-speed 5G networks enable cutting-edge research collaborations.

Case Study 5: University of British Columbia (UBC), Vancouver, Canada – UBC is extremely influenced by Vancouver's sustainability initiatives. In Vancouver, the primary aims of smart city initiatives are enhancing public transit, reducing carbon footprint, and utilizing more renewable energy sources. UBC has made all three of these goals integral to its operations. The UBC campus is a living model of a sustainable community with smart grids, district energy systems, and green transportation complementing the city's own green policies. The campus has also implemented smart waste management systems and IoT sensors to track and reduce energy and water consumption.

1.2 The Role of Education in Smart Urban Development

A potential remedy for the environmental issues brought on by fast urbanization is smart cities (Toli & Murtagh, 2020). Cities today are becoming more intelligent and prepared for the future. The ways in which cities are fostering social and financial inclusion as well as sustainable development are thus important topics. Smart universities, on the other hand, are necessary for the development of smart cities. These institutions must make significant efforts to become sustainable themselves in addition to emphasizing technology-based curriculum and infrastructure as well as community and governmental participation. For universities hoping to become smart parts of smart cities, economic and social sustainability, which is defined by effective resource utilization, budget transparency, and gender diversity, is essential (Bhatia, 2018). According to Rothaermel, Agung, and Jiang (2007), universities have historically been primarily responsible for teaching students how to undertake fundamental research and, from an ecosystem viewpoint, for producing and disseminating information to the ecosystem's stakeholders. According to Ardito, Ferraris, Messeni Petruzzelli, Bresciani, and Del Giudice (2019), universities must reconsider and modify their role in this new context due to the unique characteristics of Smart-City projects (SCPs) and the new opportunities brought about by the use of ICT in SCPs. This is especially important when considering the intricate dynamics of engagement among numerous stakeholders. Another developing trend indicates that universities are becoming more interested in new types of entrepreneurial activity, such as start-ups centred on technology generated at the university or licensing new technologies to small private businesses that demonstrate creative outcomes (Powers & McDougall, 2005).

We conducted a literature review of publications from the introduction of smart university and smart city concepts, spanning the years 1997–2024, to analyse the role of education in smart urban development. We selected the Web of Science database for the analysis. To do the analysis, we used VOSviewer. Maps of countries, publications, journals based on networks (co-citation), maps with multiple objects, and maps with a large number of publications may all be produced by this computer software. On the basis of common networks, it may also generate term maps (Rosak-Szyrocka, 2024; Rosak-Szyrocka & Knop, 2024). This software was used in our study to show nation cooperation networks, co-authorship networks, and keyword co-occurrence networks. Additionally, we depicted the structure of the scientific research landscape in smart cities and smart institutions and conducted cluster analysis. One of the main reasons for choosing VOSviewer was its ability to handle co-authorship and keyword co-occurrence data from Web of Science database. The use Web of Science as the main database guarantees that the study is grounded on reputable, peer-reviewed literature, offering trustworthy insights into the evolution of the area. The content of 1,058 research articles has been taken into consideration between 1997 and 2024 (Figure 1.1).

Articles (613), book chapters (44), proceeding papers (458), review articles (26), and easy access (11) were taken into consideration. The search documents related to the 'smart universities' and 'smart city' issue are in title, abstract, and keywords. Figure 1.2 shows the number of publications and citations from 1997 to 2024.

Figure 1.2 features three stages of the vast expanse of these fields. The first phase, covering the years 1997 to around 2005, remained fairly modest in regard to publications and citations, signifying the novelty of the concepts under exploration. The second phase covers 2006 to 2015 and is characterized by a steady increase in publications and citations. Such an increase is in conjunction with the respective developments of the ICT and IoT, as well as sustainability that are becoming increasingly important. Researchers and practitioners had begun to recognize the potential for these technologies to disrupt urban centres and educational institutions. Citation numbers in this early phase continued to gradually increase which tentatively suggests that cutting-edge lines of research

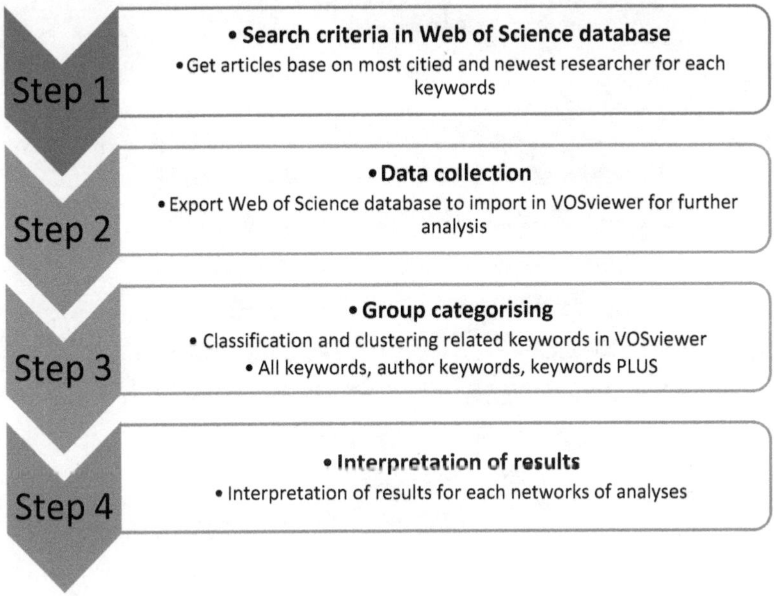

Figure 1.1 Steps in bibliometric analysis.

Source: own study.

began to inform emerging findings in ways that facilitated other lines of research and application. The last phase from 2016 until 2024 shows a notable increase in both publications and citations, which is aligned with global trends towards digital transformation and connection of smart technologies in urban and educational systems. This fast expansion is in line with global policy efforts supporting smart cities and universities around the world that are adopting smart campus solutions. This points towards the role of trans-disciplinary research to meet some of the bigger challenges facing humanity, be it the most efficient use of scarce resources, be it sustainability, or dealing with urbanization and remodelling educational infrastructure through the use of technology. Such articles aid in comprehending the relationship between two key cornerstones of the research ecosystem, that is, research output and citations. While publication activity tends to surge when a new technology or policy takes hold, cites tend to follow, mimicking the time it takes for new studies to work their way into the broader story of science. The maxima in the three citation counts indicate that certain types of work may have been particularly successful in shaping the direction of research in this area, such as foundational frameworks, comprehensive reviews, or case studies of early adopters. The trends we saw also reflect the interdisciplinary nature of the research on smart cities and smart universities, which is grounded in contributions from fields that include urban planning, computer science, sustainability studies, and educational technology. This variation has enriched the discussion, enabling the investigator to address

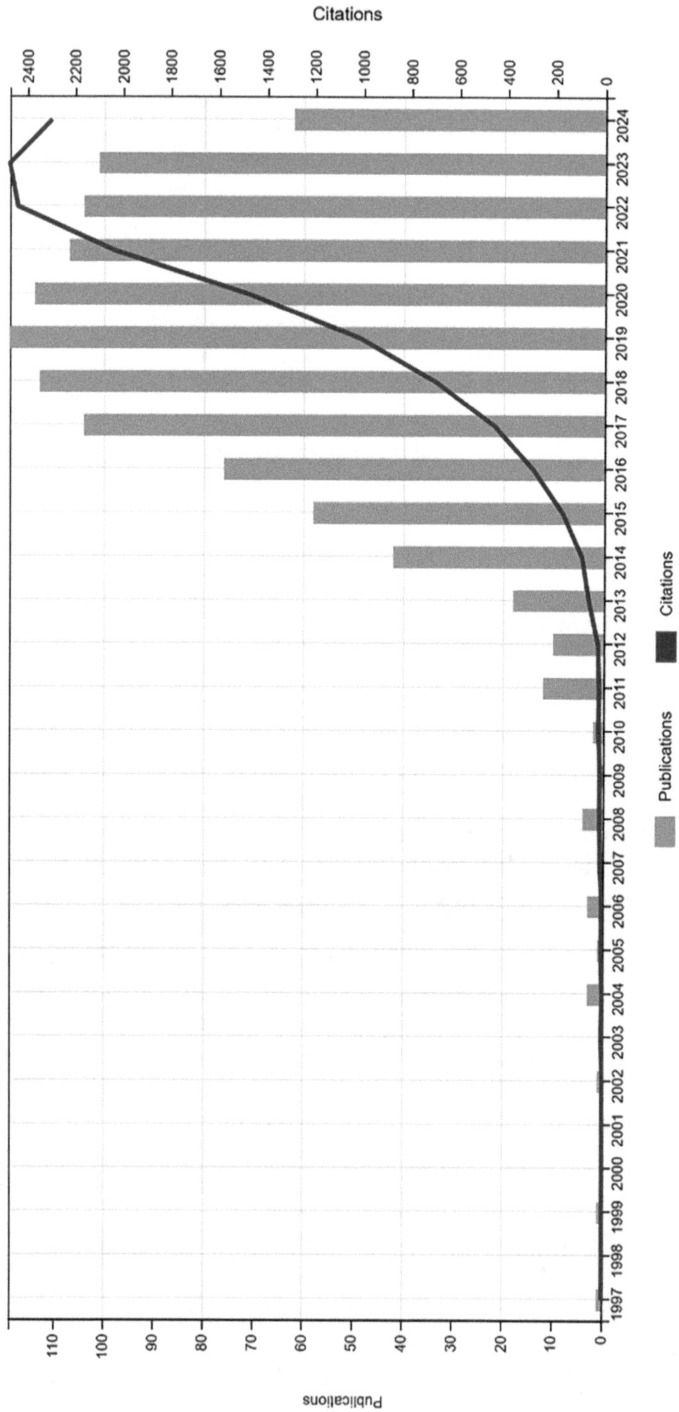

Figure 1.2 Number of publications and citations 1997–2024.

Source: own study.

complex issues with a multi-frontal approach. Also, in a technology-centred world, the interest in these areas has been increasing in the United States, particularly in natural and operational environments. The two dimensions are presented in Figure 1.2, where we identify the key stages of growth and the driving forces of scholarly activity addressed to smart cities and smart universities. Trends highlighted in the graph also illustrate the importance of networking, communicative action, and technological progress in tackling the problems of the modern world while finding sustainable urban living and learning practices. This review not only offers retrospective insight but also guides future direction for research and practice. Figure 1.3 examines how individual countries tend to drive external input into the academic conversation, focusing especially on the partnerships that root the development of smart cities and smart education systems.

Figure 1.3 highlights the importance of international collaboration when addressing challenges like urbanization, sustainability, and the integration of technology in education. Countries such as the United States, China, and European countries are hubs that fuel the generation and spread of knowledge in the field thanks to their strong research production and extensive networks. Having such partnerships facilitates the exchanging of innovative methods as well as best practices and tools that are important for intelligent urban design and education reform. The analysis also shows the variety of approaches

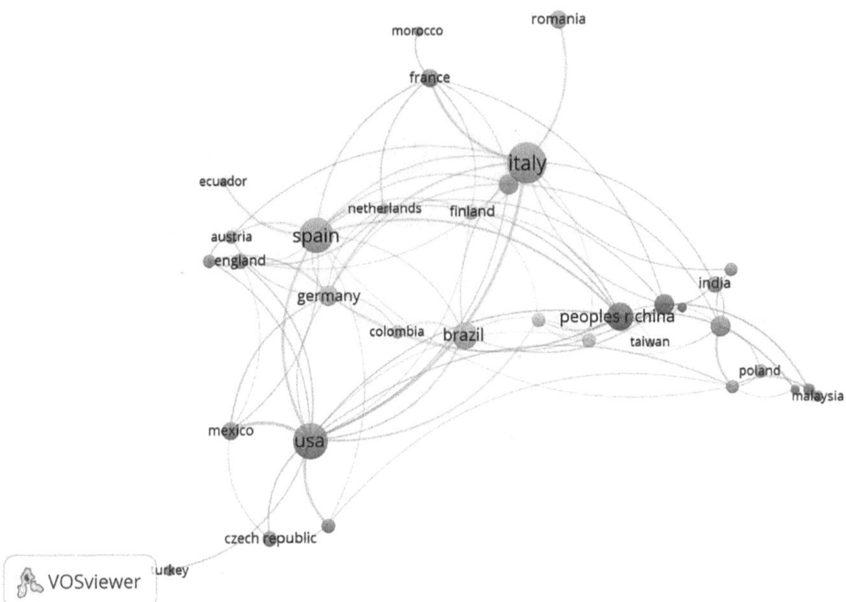

Figure 1.3 Co-authorship countries (minimum number of occurrence: 5).

Source: own study.

reflected by regional priorities and technological capabilities. High technology in developed nations usually pioneers the initial technology but developing nations are the ones providing the enabling infrastructure or the testbeds that adapt those technologies to the local context. This power plays a vital role in establishing strong and enduring smart city and smart campus projects, which can be used internationally. The network of co-authorship also shows how education is a key factor in promoting collaboration. Universities and research institutions stand at the heart of internationalization and examine global academic networks in advancing research into smart cities. Not only do these partnerships improve the educational infrastructure; by introducing concepts such as the IoT, AI, and renewable energy, they also bolster urban development objectives. Transferring data, tools, and methodologies across national borders is vital to speed up innovation, and make sure that lessons learned in one country can be adapted to others. Furthermore, the number fits more broadly and should be part of the wider story of education as an engine for smart, urban growth. The collaboration enables the framework to guide the innovative resolution of the cross-cutting issues of urbanization including, but not limited to, resource management, mobility, and governance. Universities play an essential role in this and benefit from having global networks to inspire world-class curricula and insightful research. Figure 1.4 shows the co-occurrence of all

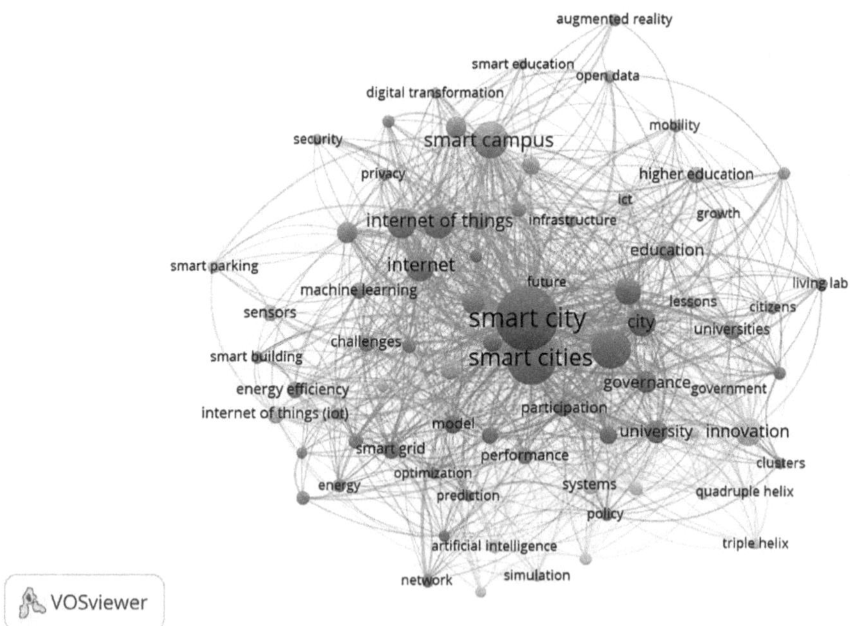

Figure 1.4 Co-occurrence all keywords (minimum number of occurrence: 5).

Source: own study.

keywords and provides important insights into the thematic structure and trends in research related to smart cities and smart universities.

A dominant node at the centre of this network is 'smart city'. Around it is another, 'smart cities', which serves to anchor the whole research discourse. Around those several key concepts are interconnected themes: 'education', 'smart campus', 'universities', and 'innovation'. This not only testifies to the critical role of education in the development and implementation of smart urban systems but also demonstrates just how complex an idea 'smart city' really is. The closeness and dense connection between 'smart campus' achievement and 'education' suggest that educational institutions serve as the new forces for driving innovation in this aspect of urban development. These campuses are portrayed as microcosms of smart cities, where new technologies as well as policies are tested and refined. IoT, AI, and a digital transformation now appearing on campuses not only make infrastructure more efficient but also embody a space for students, researchers, and faculty members to collaborate in the creation of practical responses to urban problems. Words in the green cluster, such as 'infrastructure', 'mobility', 'growth', and 'citizens', reveal the convergence between smart campuses and municipal sustainability goals. This connection underscores that through education we have the tools, models, and frameworks needed to take solutions from campuses to larger settings, such as cities. The universities are not merely providing technology. As the red cluster shows, education also determines how things will be done through participation, policy, and government. Instead of merely producing new knowledge or providing policy advice, universities are now actively involved in the intricacies of urban life. They actively involve citizens, regulators, and communities in hands-on work to build up urban solutions involving education. This interactive process becomes a force for city development. The network demonstrates the sweeping nature of technological innovation in education. As 'Internet of Things', 'artificial intelligence', 'machine learning', and 'augmented reality' become part of the 'smart campus', we perceive a broad connection between them and education. These technologies will revolutionize teaching and learning processes. They also bring campus facilities into being— 'smart' enough that they can monitor and optimize them. Leverages of these capabilities spread around the campus, making resource use more efficient in real time and reducing environmental impacts overall. It creates environments able to meet the needs of all stakeholders. Innovations lead to wider strategies for towns and cities, showing that the concept of a smart campus can be redeployed to fashion an intelligent city. Visible connection between the words 'sustainability', 'energy efficiency', 'optimization', and 'performance' underlies that education contributes to the environmental conception of moving forward urban development. Universities are taking the lead in creating resource-efficient campuses: clean energy, energy saving, and minimizing waste. This cooperative mode of development implies aligning the latest in technology with what society needs and what best serves urban planning goals.

Education is not only a kind of practice but also serves as something driving the Web of development in smart cities. As a modern campus, it is a test field to train urban strategies for dealing with problems that crop up now and then that will be lacking in changed cities in the future. It might, furthermore, even be able to provide examples that could serve as models on which other smart cities around the world will be founded. Universities foster knowledge, innovation, and collaboration. They make it possible for cities to become intelligent, multiplicative, and sustainable urban environments. Keywords in the network are interconnected figuratively. Figure 1.4. shows that education lays a crucial foundation for building resilient, sustainable urban systems with a long-term vision of progress and development. Figure 1.5 shows the co-occurrence author keywords.

In Figure 1.5, special mention deserves the central role of education in shaping cities that are sustainable, technologically advanced, and inclusive. Smart city emerges at the core in the figure, emphasizing that a holistic, wide-ranging urban environment can be treated as one. The key factor in initiating such transformation is to be found in education itself. Overarching nodes, such as 'smart university', 'smart campus', 'higher education', and 'smart education', indicate that educational institutions are at the centre. They are centres for research, technology development, and transmission of knowledge, the key to solving urban problems. Contacts between education-related nodes and technical concepts, such as IoT, machine learning, AI, and distributed computing technologies, show universities and campuses have a strong driving effect on technology innovations.

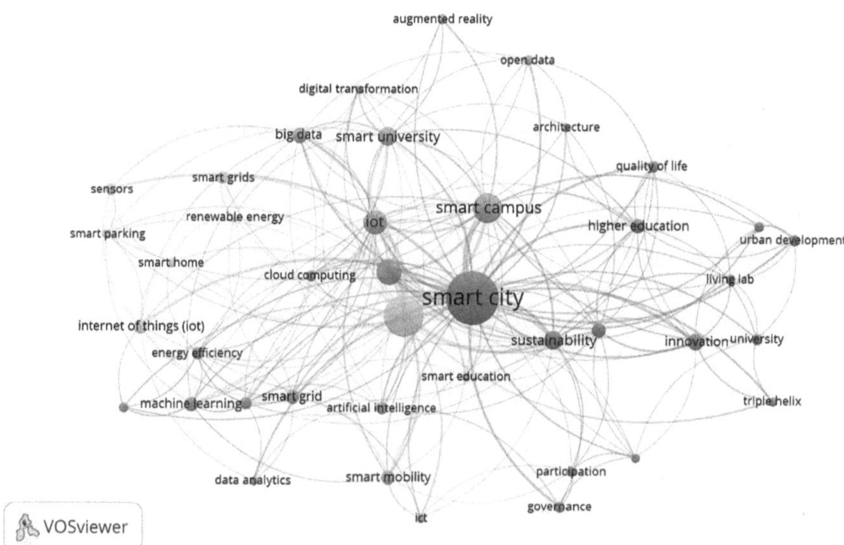

Figure 1.5 Co-occurrence author keywords (minimum number of occurrence: 5).

Source: own study.

Smart universities and smart campuses are platforms on which to test new technologies, and methods available for improvement of urban infrastructure with more efficiency in energy use and less pollution (as in intelligent heat/electricity systems). By equipping citizens with technical and managerial skills, educational institutions make sure that smart cities have the necessary human resources to sustain the rapid evolution of technology and new urban development. Sustainability is another major area relating to education in smart urban development. By researching replacement energy sources, how to save fuel, and at-smart systems for the power network, universities contribute to make cities sustainable. More than just making a difference to urban living, these efforts also help cultivate novel methods of environmental control (e.g., reducing what's consumed and the construction of eco-friendly urban systems). Linkages between higher education, sustainability, and the quality of urban life tell us that education plays a transformative role in raising peoples' level of material existence, forming broad-minded citizens, and even inducing environmental awareness.

Figure 1.5 also mirrors the collaborative nature of education in smart cities. Nodes like 'triple helix' and 'living labs' appear to indicate that educational institutions are vital to promoting cooperation among government, the private sector, and academia for urban innovative practice. Such cooperative endeavours ensure educational initiatives will complement urban development at large and be of service to the spread of technology in cities, creating an ecosystem for the realization of a smart city. And higher education's connection with quality of life and innovation reflects its role in making urban environments more liveable, equitable, and dynamic. In VOSviewer, Keywords Plus, by clustering terms that tend to occur simultaneously in multiple documents, can show emerging thematic trends and relationships (Figure 1.6).

Figure 1.6 draws attention to the roles that higher learning plays as a neutral platform between technology integration and innovative technology breakthroughs. It even serves as a guiding force, one of which cannot exist without another. Figure 1.6 marks off the importance that education has given to smart city rising economic centres. It highlights such institutions in addition to acting as research centres' role as facilitators of urban design and construction. At universities, educational technological research can supply predictions about resource management challenges in cities and while privacy issues are taken care of by a system architecture to confine the damage from leaking incidents.

1.3 Global Trends in Smart Universities and Smart Cities

Our biggest challenge in this new century
is to take an idea that seems abstract
– sustainable development –

and turn it into a reality for all the world's people.
–Kofi Annan, Former United Nations Secretary-General

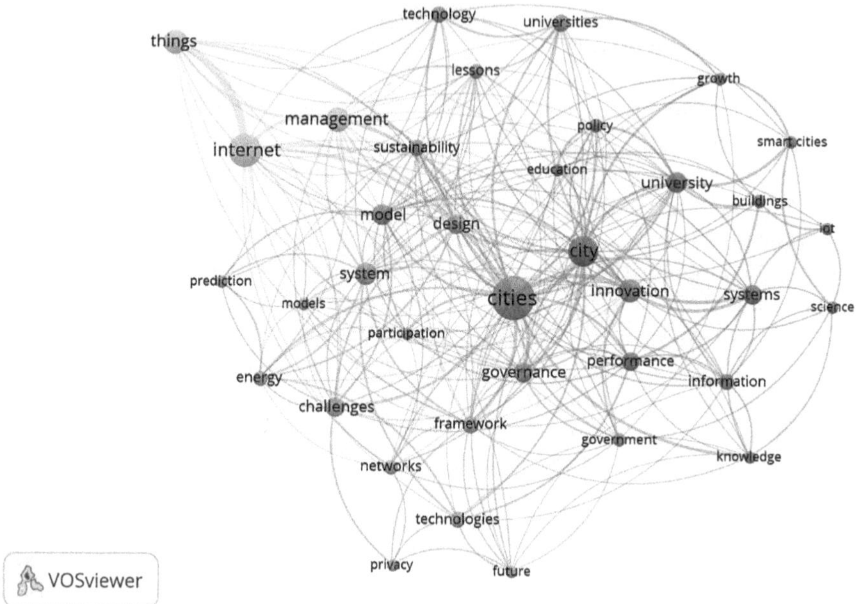

Figure 1.6 Co-occurrence Keywords Plus (minimum number of occurrence: 5).

Source: own study.

With digital technology and urbanization rapidly developing, cities have been transformed into 'smart cities'. They now contain data-driven solutions aimed at improving everywhere from public utilities to sustainability and citizen welfare. In the same vein, traditional educational institutions are now 'smart universities', using digital tools and technology to increase both research quality and the impact it has on society as well as students' learning experiences. Underground networked sensors, both fixed and movable, as well as grouped high-tech applications enable these cities main driving force for urban development. When it comes to real-time monitoring over such a wide area, there are a variety of means and above all the use of IoT has made this possible. That means all the water and electric power used by every part of a city; not only are gas turbine emissions monitored but also traffic movements in real time (Rabbi Shawon & Rahman Chowdhury, 2013). New urban examples show how cities can use public–private partnerships to bring about technology-driven solutions favoured by efficient growth policy goals. With the smart city growing in strength, universities have taken advantage of these feats. Techniques are now being developed that use technology to redesign study environments, optimize the running of university campuses, and extend the universities' role in research. Meanwhile, making full use of digital transformation, IoT networking, and sustainable energy practices has become a popular topic on university campuses. These are among the ingredients for education innovation in smart universities (Le Phuc Do, Viet, Mai, & Trinh, 2024). Vietnamese universities are using technology in architectural and urban design programs to train future urban planners who can create inclusive planning for Smart Cities (Le Phuc Do et al., 2024). This also reflects changing curricular emphases in higher education shifting from single-discipline studies to technology in education, sustainable development studies, and city planning. In Table 1.4, key trends connecting smart cities and smart universities are presented.

Table 1.4 Key trends connecting smart cities and smart universities

Key trend	*Description*
Technology integration for innovation	Smart cities and smart universities rely on technological technology-based trends. Infrastructure, energy optimization, and real-time communication systems (Holderness et al., 2014) exhibit the adoption of IoT, big data, and geospatial tools in infrastructure planning. For example, universities are using geosocial intelligence to address critical problems like building infrastructure resilience in cities at high risk of natural disasters.
Renewable energy and sustainability integration	One of the major trends we have in general in smart city and smart university development is sustainability. To this end, cities are implementing hybrid renewable energy solutions that will decrease carbon emissions, increase energy efficiency, as well as help meet climate goals. At the same time, intelligent universities also embed sustainability into campus management and in the construction of curricula to provide energy-efficient systems for students, increasing their awareness of sustainability (Şekeroğlu & Erol, 2022). Poland also emphasizes the relationship between demography and ecology through consumption behaviors, while illustrating how universities can engage the communities they focus on in pursuit of urban sustainability ('Typology of Urban Silver Singles in Selected Cities in Poland', 2024). Such efforts complement the wider United Nations Sustainable Development Goals (SDGs)
The integration of smart governance practices in both cities and universities	Smart cities are shifting towards data-driven governance to improve decision-making, transparency, and citizen participation. Universities play a key role in developing and testing frameworks for urban governance, leveraging predictive models and data analysis to optimize public services and infrastructure (Rabbi Shawon & Rahman Chowdhury, 2013). The integration of smart governance practices in both cities and universities demonstrates the importance of policy alignment and participatory decision-making. For example, research highlights how universities contribute to the design of governance models that enable real-time data collection and citizen-centric urban management.

(*Continued*)

Table 1.4 (Continued)

Key trend	*Description*
Smart campus infrastructure and digital learning platforms	A trend towards smart universities is the establishment of smart campus infrastructure that is similar in technology to smart cities. Bigger education institutions are already adopting IoT-based systems such as smart classrooms, automatic energy management, real-time campus monitoring, etc., to improve operational efficiency and student experience. AI applications such as cloud connectivity and IoT devices can be quickly implemented and tested on smart campuses, serving as a hotbed for the innovative technologies heading into smart cities. AI and cloud computing are also integrated with digital learning platforms to help deliver personalized learning experiences and foster flexibility and accessibility in education (Hoke, 2025). The role of universities includes enabling smart infrastructure that can make a large impact on society and working in partnership with academic institutions, industries, and municipalities. This partnership leads to solutions that can be applied to real-life challenges in the cities of how they can optimize traffic, manage water, and reduce waste.
Public private partnership (PPP) for innovation	PPPs are a growing foundation for technological and social innovation in smart cities and smart universities. Such partnerships also provide opportunities for academic institutions to partner with industry, governments, and technology providers to deploy urban solutions. Universities are the backbone of research and development, while private organizations bring the tech know-how and funding to scale innovation. For instance, PPPs are one of the prominent mechanisms to develop renewable energy, smart mobility, and digital infrastructure projects in cities and campuses (Moolngearn & Kraiwanit, 2024). Moreover, PPP whereby both private and public sectors address smart universities accelerates practical applications and technological solutions for smart city ecosystems alongside improvement in innovation and sustainability.
AI and automation	AI and automation are radically transforming civic oversight and pedagogy in modern municipalities and institutions of higher education. Innovative cities have taken action by taking advantage of predictive analytics capabilities (including directing traffic, conserving energy, and preparing for calamity) to make the efficient deployment of resources viable. At the same time, universities began the adoption of AI to support both instruction and scholarship and administrative processes. AI-powered platforms now provide bespoke pathways for learning, streamlining how course materials are presented and digitizing management processes on campus such as security and facility upkeep.

Source: own study.

Smart universities are at the centre of the growth and success of smart cities. They act as engines of innovation, generating the research, technology, and talent needed to drive urban development. As such, universities are incorporating interdisciplinary education programs to train students for careers in urban planning, data science, and energy management. An important feature of this phenomenon is the shift to predictive models and privacy frameworks in response to new challenges in the urban area (Holderness et al., 2014). In response to this challenge, universities or research-led projects are emerging that attempt to provide hardware and software focused on frameworks for secure and ethical innovation for big data in the smart city context. In addition to this, the universities function as the interface of citizens engagement and knowledge dissemination. Programs for educating leaders focus on participatory governance, which trains the next generation of leaders who will draw in stakeholders and close divides between societies and technologies. This follows the exercise of implementation of citizen-centric policies for inclusive urban development by cities.

References

Alghamdi, A., & Shetty, S. (2016). Survey toward a smart campus using the internet of Things. In *2016 IEEE 4th international conference on future Internet of Things and Cloud (FiCloud)* (pp. 235–239). IEEE.

Alizadeh, T. (2017). An investigation of IBM's smarter cites challenge: What do participating cities want? *Cities*, *63*, 70–80.

Alvarez-Campana, M., López, G., Vázquez, E., Villagrá, V. A., & Berrocal, J. (2017). Smart CEI Moncloa: An IoT-based platform for people flow and environmental monitoring on a smart university campus. *Sensors*, *17*(12), 2856, https://www.mdpi.com/1424-8220/17/12/2856

Ardito, L., Ferraris, A., Messeni Petruzzelli, A., Bresciani, S., & Del Giudice, M. (2019). The role of universities in the knowledge management of smart city projects. *Technological Forecasting and Social Change*, *142*, 312–321, https://www.sciencedirect.com/science/article/pii/s0040162517317754

Bhatia, S. (2018). Sustainable smart universities for smart cities. *Journal of Economics, Management and Trade*, *21*(12), 1–11.

Cata, M. (2015). Smart university, a new concept in the Internet of Things. In *2015 14th RoEduNet International conference - networking in education and research (RoEduNet NER)*. IEEE.

Chen, H., Jiang, W., Yang, Y., Yang, Y., & Man, X. (2017). State of the art on food waste research: A bibliometrics study from 1997 to 2014. *Journal of Cleaner Production*, *140*, 840–846.

Chuling, L., Xie, Z., & Peng, P. (2009). A discussion on the framework of smarter campus. In *2009 Third international symposium on intelligent information technology application* (pp. 479–482). IEEE.

Coccoli, M., Guercio, A., Maresca, P., & Stanganelli, L. (2014). Smarter universities: A vision for the fast changing digital era. *Journal of Visual Languages & Computing*, *25*(6), 1003–1011, https://www.sciencedirect.com/science/article/pii/s1045926x140086x?casa_token=cllj6rexp48aaaaa:duswcycqbb607fvoyt2x3nkog1_ckxe2np_3wqqpnbf0ztbhdrnakpfrcwfpj2sqefpwympjify

El Khatib, M., Alhammadi, M., & Almulla, A. (2024). Attributes of SMART education in SMART cities. *International Journal of Business Analytics and Security (IJBAS)*, *4*(2), 157–178, https://journals.gaftim.com/index.php/ijbas/article/view/359

Falconer, G., & Mitchell, S. (2012). *Smart city framework a systematic process for enabling smart+connected communities*. Cisco Internet Business Solutions Group (IBSG).

Fortes, S., Santoyo-Ramón, J. A., Palacios, D., Baena, E., Mora-García, R., Medina, M., et al. (2019). The campus as a smart city: University of Málaga environmental, learning, and research approaches. *Sensors (Basel, Switzerland)*, *19*(6).

Gao, X., Sun, Y., Hao, L., Yang, H., Chen, Y., & Xiang, C. (2018). A new soft pneumatic elbow pad for joint assistance with application to smart campus. *IEEE Access*, *6*, 38967–38976.

Gascó-Hernandez, Mila (2018). Building a smart city: Lessons from Barcelona. *Communications of the ACM*, *61*(4), 50–57.

Halawani, T., & Mohandes, M. (2003). Smart card for smart campus: KFUPM case study. In *10th IEEE international conference on electronics, circuits and systems, 2003. ICECS 2003. Proceedings of the 2003* (pp. 1252–1255). IEEE.

Hoke, T. (2025). *AI with a heart: Integrating emotional intelligence and sustainability in smart education for 21st century learners* (pp. 458–465). Singapore: Springer.

Holderness, T., Michael, K. & Turpin, E. (2014). Infrastructure planning through geosocial intelligence: Using twitter as a platform for rapid assessment and civic co-management during flooding in Jakarta. In *Defining the sensor society conference*. Queensland: University of Queensland.

Hollands, R. G. (2020). Will the real smart city please stand up?: Intelligent, progressive or entrepreneurial? In *The Routledge companion to smart cities* (pp. 179–199). Routledge.

IEEE 5th International forum on Research and Technology for Society and Industry (RTSI) (2019). IEEE.

International Conference on Information Technology Systems and Innovation (ICITSI) (2017). IEEE.

Kwok, L. (2015). A vision for the development of i-campus. *Smart Learning Environments*, *2*(1), 1–12, https://link.springer.com/article/10.1186/s40561-015-0009-8

Le Phuc Do, T., Viet, L. Q., Mai, T. T. Q., & Trinh, T. A. (2024). The prospect of applying technology in university education: The field of architectural and urban design inclusive smart city in Vietnam. *The 2nd International Conference: Resiliency by Technology and Design (RTD 2024)*, *186*, 511–526, https://www.atlantis-press.com/proceedings/rtd-24/126005336

Lin, Y.-B., Chen, L.-K., Shieh, M.-Z., Lin, Y.-W., & Yen, T.-H. (2018). CampusTalk: IoT devices and their interesting features on campus applications. *IEEE Access*, *6*, 26036–26046.

Liu, D., Huang, R., & Wosinski, M. (2017). *Smart learning in smart cities*. Springer.

Majeed, A., & Ali, M. (2018). How Internet-of-Things (IoT) making the university campuses smart? QA higher education (QAHE) perspective. In *2018 IEEE 8th annual computing and communication workshop and conference (CCWC)* (pp. 646–648). IEEE.

Malatji, E. M. (2017). The development of a smart campus - African universities point of view. In *2017 8th international renewable energy congress (IREC)* (pp. 1–5). IEEE.

Manager (2018). *The weight of cities*. https://www.resourcepanel.org/reports/weight-cities

Manqele, L., Dlodlo, M., Coetzee, L., Williams, Q., & Sibiya, G. (2015). Preference-based Internet of Things dynamic service selection for smart campus. In *AFRICON 2015* (pp. 1–5). IEEE.

Manufacturing & Industrial Engineering Symposium (MIES) (2016). IEEE.

Mattoni, B., Pagliaro, F., Corona, G., Ponzo, V., Bisegna, F., Gugliermetti, F., & Quintero-Nunez, M. (2016). A matrix approach to identify and choose efficient strategies to develop the Smart Campus. In *2016 IEEE 16th international conference on environment and electrical engineering (EEEIC)*. IEEE.

Moolngearn, P., & Kraiwanit, T. 2024. Implication of smart economy governance: A perspective of smart cities in an emerging country. *Governance and Regulation*, 13(2), 431–442.

Moura, P., Moreno, J. I., López López, G., & Alvarez-Campana, M. (2021). IoT platform for energy sustainability in university campuses. *Sensors*, *21*(2).

New report: Building energy-smart cities is fastest, easiest way to address indian energy crisis (2012, November 01). Retrieved December 04, 2024, https://www.nrdc.org/press-releases/new-report-building-energy-smart-cities-fastest-easiest-way-address-indian-energy

Nie, X. (2013). Research on smart campus based on cloud computing and Internet of Things. *Applied Mechanics and Materials*, *380–384*, 1951–1954, https://www.scientific.net/amm.380-384.1951

O'Grady, M., & O'Hare, G. (2012). Computer science. How smart is your city? *Science (New York, N.Y.)*, *335*(6076), 1581–1582.

Ojo, A., Curry, E., & Zeleti, F. A. (2015). A tale of open data innovations in five smart cities. In *2015 48th Hawaii international conference on system sciences* (pp. 2326–2335). IEEE.

Omotayo, T., Awuzie, B., Ajayi, S., Moghayedi, A., & Oyeyipo, O. (2021). A systems thinking model for transitioning smart campuses to cities. *Frontiers in Built Environment*, *7*, 755424.

Omotayo, T., Moghayedi, A., Awuzie, B., & Ajayi, S. (2021). Infrastructure elements for smart campuses: A bibliometric analysis. *Sustainability*, *13*(14), 7960.

Omotayo, T., Olanipekun, A., Obi, L., & Boateng, P. (2020). A systems thinking approach for incremental reduction of non-physical waste. *Built Environment Project and Asset Management*, *10*(4), 509–528.

Owoc, M., & Marciniak, K. (2013). Knowledge management as foundation of smart university. In *2013 Federated conference on computer science and information systems* (pp. 1267–1272). Krakow, Poland

Pagliaro, F., Mattoni, B., Gugliermenti, F., Bisegna, F., Azzaro, B., Tomei, F., & Catucci, S. (2016). A roadmap toward the development of Sapienza Smart Campus. In *2016 IEEE 16th international conference on environment and electrical engineering (EEEIC)*. IEEE.

Pan, M., Zhao, X., Lv, K., Rosak-Szyrocka, J., Mentel, G., & Truskolaski, T. (2023). Internet development and carbon emission-reduction in the era of digitalization: Where will resource-based cities go? *Resources Policy*, *81*, 103345, https://www.sciencedirect.com/science/article/pii/s0301420723000533

Pinto, L. G. P., Romano, R. R., & Tomoto, M. A. (2017). From the university to smart cities—How engineers can construct better cities in BRIC's countries: A real case from smart campus FACENS. *Advances in The Human Side of Service Engineering*, *494*, 347–354, https://link.springer.com/chapter/10.1007/978-3-319-41947-3_32

Powers, J. B., & McDougall, P. P. (2005). University start-up formation and technology licensing with firms that go public: A resource-based view of academic entrepreneurship. *Journal of Business Venturing*, *20*(3), 291–311, https://www.sciencedirect.com/science/article/pii/s0883902604000291

Rabbi Shawon, R. E., & Rahman Chowdhury, M. S. (2013). *Transforming urban living in the USA: The role of IoT in developing smart cities.*

Rico-Bautista, D., Guerrero, C. D., Collazos, C. A., Maestre-Góngora, G., Hurtado-Alegría, J. A., Medina-Cárdenas, Y., & Swaminathan, J. (2021). Smart University: A vision of technology adoption. *Revista Colombiana de Computación*, *22*(1), 44–55, https://revistas.unab.edu.co/index.php/rcc/article/view/4153

Rosak-Szyrocka, J. (2024). The role of artificial intelligence in digital education. Scientific Papers of Silesian University of Technology, Organization and Management Series No. 195, Silesian University of Technology Publishing House.

Rosak-Szyrocka, J., Apostu, S. A., & Pupp, Z. (2024). The relationship between technology of higher education in the case of European countries, a panel approach. *Polish Journal of Environmental Studies, 33*(3), 2813–2821.
Rosak-Szyrocka, J., & Knop, K. (2024). Navigating the Fourth Industrial Revolution: Insights from a comprehensive bibliometric study on Industry 4.0. *Production Engineering Archives, 30*(4), 501–519.
Rosak-Szyrocka, J., & Tiwari, S. (2023). Structural Equation Modeling (SEM) to test sustainable development in University 4.0 in the ultra-smart society era. *Sustainability, 15*(23), 16167, https://www.mdpi.com/2071-1050/15/23/16167
Rosak-Szyrocka, J., Zywiolek, J., Zaborski, A., Chowdhury, S., & Hu, Y.-C. (2022). Digitalization of higher education around the globe during Covid-19. *IEEE Access, 10*, 59782–59791.
Rothaermel, F. T., Agung, S. D., & Jiang, L. (2007). University entrepreneurship: A taxonomy of the literature. *Industrial and Corporate Change, 16*(4), 691–791.
Sari, M. W., Ciptadi, P. W., & Hardyanto, R. H. (2017). Study of smart campus development using Internet of Things technology. *IOP Conference Series: Materials Science and Engineering, 190*(1), 12032, https://iopscience.iop.org/article/10.1088/1757-899x/190/1/012032/meta
Schaffers, H., Komninos, N., Pallot, M., Trousse, B., Nilsson, M., & Oliveira, A. (2011). Smart cities and the future internet: Towards cooperation frameworks for open innovation. In *Lecture notes in computer science* (pp. 431–446). Berlin, Heidelberg: Springer Berlin Heidelberg.
Şekeroğlu, A., Erol, D. (2022). Methodology for spatial analysis of hybrid renewable energy potential. *Megaron*, 17(3), 511–525.
Smart cities study: International study on the situation of ICT, innovation and knowledge in cities (2012). https://issuu.com/uclg_cdc/docs/en-smart_cities_study
Smart cities: Digital solutions for a more livable future (2018). https://www.mckinsey.com/capabilities/operations/our-insights/smart-cities-digital-solutions-for-a-more-livable-future
Smart Education and Learning Market (2024). *Global report [2027]*. Retrieved December 03, 2024, https://www.fortunebusinessinsights.com/industry-reports/smart-education-and-learning-market-101942
Sotres, P., Santana, J. R., Sanchez, L., Lanza, J., & Munoz, L. (2017). Practical lessons from the deployment and management of a smart city internet-of-things infrastructure: The SmartSantander Testbed case. *IEEE Access, 5*, 14309–14322.
Sustainable, resource efficient cities - Making it happen. Department of Economic and Social Affairs (2024). Retrieved December 04, 2024, https://sdgs.un.org/publications/sustainable-resource-efficient-cities-making-it-happen-17747
The smart city vision: How innovation and ICT can build smart, "livable", sustainable cities (2010).
Toli, A. M., & Murtagh, N. (2020). The concept of sustainability in smart city definitions. *Frontiers in Built Environment, 6*, 1–10.
Tsai, I.-C., & Yeh, C.-H. (2015). Integrating SERVQUAL and importance-performance analysis for assessing smart campus service quality: A case study of an english training programme in Vietnam. In *2015 IEEE 15th international conference on advanced learning technologies* (pp. 436–440). IEEE.
Typology of urban silver singles in selected cities in Poland (2024).
UN Environment. (2019, March 7). UN Environment 2018 Annual Report. United Nations Environment Programme. Retrieved April 20, 2025, from https://www.unep.org/resources/un-environment-2018-annual-report
Verstaevel, N., Boes, J., & Gleizes, M.-P. (2017). From smart campus to smart cities issues of the smart revolution. In *2017 IEEE smartworld, ubiquitous intelligence & computing, advanced & trusted computed, scalable computing & communications,*

cloud & big data computing, internet of people and smart city innovation (SmartWorld/SCALCOM/UIC/ATC/CBDCom/IOP/SCI) (pp. 1–6). IEEE.

Villegas-Ch, W., Palacios-Pacheco, X., & Luján-Mora, S. (2019). Application of a smart city model to a traditional university campus with a big data architecture: A sustainable smart campus. *Sustainability*, *11*(10), 2857, https://www.mdpi.com/2071-1050/11/10/2857

Washburn, D., Sindhu, U., with Balaouras, S., Dines, R. A., Hayes, N. M. & Nelson, L. E. (2010). Helping CIOs Understand "Smart City" Initiatives by Doug Washburn and Usman Sindhu for CIOs, Making Leaders Successful Every Day, https://www.forrester.com/report/Helping-CIOs-Understand-SmartCity-Initiatives/RES55590

WebWire (2024). *IBM offers smarter city assessment tool to help cities prepare for challenges and opportunities of unprecedented urbanization*. Retrieved December 04, 2024, https://www.webwire.com/ViewPressRel.asp?aId=97891

Zanella, A., Bui, N., Castellani, A., Vangelista, L., & Zorzi, M. (2014). Internet of Things for smart cities. *IEEE Internet of Things Journal*, *1*(1), 22–32.

Zhai, X., Dong, Y., & Yuan, J. (2018). Investigating learners' technology engagement - A perspective from ubiquitous game-based learning in smart campus. *IEEE Access*, *6*, 10279–10287.

Zhang, W., Zhang, X., & Shi, H. (2018). MMCSACC: A multi-source multimedia conference system assisted by cloud computing for smart campus. *IEEE Access*, *6*, 35879–35889.

Zhang, Y., Dong, Z. Y., Yip, C., & Swift, S. (2020). Smart campus: A user case study in Hong Kong. *IET Smart Cities*, *2*(3), 146–154.

Zygiaris, S. (2013). Smart city reference model: Assisting planners to conceptualize the building of smart city innovation ecosystems. *Journal of the Knowledge Economy*, *4*(2), 217–231.

2 The Role of Technology in Smart Universities

2.1 Internet of Things (IoT) on Campus

The idea behind 'smart campuses' is to upgrade various campus management-related tasks by using the Internet of Things (IoT)-enabled infrastructure. The IoT technology creates intelligent, interconnected systems that maximize resource use, boost operational efficiency, and enhance the overall campus experience (Gubbi, Buyya, Marusic, & Palaniswami, 2013). According to Atzori, Iera, and Morabito (2010), the smart campus works by monitoring various operating factors, such as energy consumption, building occupancy, and environmental conditions, enabled by IoT devices, sensors, and networks providing real-time data about the campus. This data can then be used to provide automated processes and informed decision-making which can lead to better resource management and savings (Atzori et al., 2010). IoT-enabled systems improve a building's lighting, heating, and cooling in response to occupancy and environmental systems. This leads to conserving the amount of energy consumed by the organization and its carbon footprint which is the energy-efficient model of the organization, thus saving money (Gubbi et al., 2013). Significant research has been devoted over the past years to implementing information technologies (ITs) for measuring space use in the real time, primarily for operational applications. At the same time, however, data on usage of spaces, user satisfaction, or energy consumption can also be used to inform several campus management decisions. These include strategic (long-term perspective) decisions, for example, building renovation, construction, and disposal, and tactical decisions (mid-term perspective), for example, optimization of cleaning schedules, building opening hours, and purchase quantities of food facilities. Information systems are used to improve organizational processes and provide organizational benefits. Smart campus tools collectively are addressed under the umbrella of campus information systems: they are installed to gather data, which are subsequently utilized not only in operational but also in strategic processes, with the aim of supporting universities in facing their new challenges.

Smart campus tools are more often referred to as IoT in scientific research and in practice. Gubbi et al. (2013) define IoT as connecting with sensing and actuating devices to enable the sharing of information across platforms using

DOI: 10.4324/9781003621829-3

a common framework, creating a common operating picture for supporting new applications. This is done with seamless ubiquitous sensing, data analytics, and information representation with cloud computing as unifying fabric. This definition applies to all smart campus tools but specifically focuses on the sensing and actuating devices, their output, as well as the applications that follow. Thus, smart campus is a sub-category of the IoT which extends to IoT applications.

The IoT offers a platform that enables remote device connectivity, sensing, and control via a network infrastructure (Hossain, Das, & Rashed, 2019). Because it enables the connecting of individually identifiable computer devices inside the current Internet infrastructure, IoT has grown in popularity. IoT has moved past its infancy and is poised to turn the current static Internet into a fully integrated Future Internet, driven by the widespread use of devices made possible by open wireless technology like Bluetooth, radio frequency identification (RFID), Wi-Fi, and telephonic data services, as well as embedded sensor and actuator nodes. The move towards IoT is further reflected in education, with statutory technical training bodies introducing the course as mandatory for engineering education (Malhotra, Kumar, & Dutta, 2021). IoT altered the working environments of the manufacturing industry, financial sector, entertainment, consumer services, and education. This newly introduced technology not only required amendments in the teaching tools and equipment used but also led to the introduction of new courses (Kortuem, Bandara, Smith, Richards, & Petre, 2013). IoT can deliver big data around usage patterns; the better understanding the space utilization, the more efficient utilization of existing space, and for real estate managers, better decisions on long-term demand. Moreover, space utilization data can be enriched with user feedback data and environmental variables, for example, noise levels and luminance (Valks, Arkesteijn, Koutamanis, & Heijer, 2021).

Public buildings are a prime example of smart management systems. These organizations must not only serve as models for sustainable development but also positively influence the cities and areas where they are located. The relevance of these smart building systems has increased in recent years due to the development of digitization. Buildings comprise the physical infrastructure of a campus. The systems leverage the IoT and its linked sensors to enhance or streamline building operations and user experience, among other things (Ahmed et al., 2022; Polin, Yigitcanlar, Limb, & Washington, 2023). Because of the COVID-19 epidemic, data like real-time audience counts, movement patterns, and air quality at indoor or outdoor events have become more important than ever. Specifically, low-cost sensor technology may be used to help gather a variety of data and information at a reasonable cost. This comprises environmental, human, and vehicle data, such as light or noise levels as well as climate data, such as temperature, humidity, or air quality. Visitor, space, energy, and parking management are some of the most popular uses for smart building or campus systems (Ahmed et al., 2022; An, Wang, & Chen, 2022; Monti et al., 2022). Operators can better schedule events and manage resources with the use of such technologies. These possible systems provide pertinent information about the location, resources offered, and facility occupancy to visitors. But creating these campus or smart building systems is not without its problems, such as privacy concerns, security threats, and technological hurdles (Sneesl, Jusoh, Jabar, & Abdullah, 2022). Monitoring people at building entrances is not enough to provide useful information on the distribution of visitors in bigger building complexes. Because of the high technological requirements, acceptability problems, and possible rights breaches, the deployment of full camera monitoring is often seen as inefficient (Huang, Rodriguez, Whetstone, & Habel, 2019; Sutjarittham, Habibi Gharakheili, Kanhere, & Sivaraman, 2018).

The number of electronic devices that students and campus residents use and link to the university network has increased dramatically, giving rise to vast amounts of data that may be used for decision-making (Valks et al., 2021). The potential of smart cities via the integration of IoT devices and smart technology has encouraged colleges to implement the idea on a number of campuses (Min-Allah & Alrashed, 2020). The idea of smart campuses and smart cities is still developing, however. Data security, authentication, illegal access, and sustainability are among the security issues that persist despite their enormous potential (Bhushan et al., 2020; Haque, Bhushan, & Dhiman, 2022). University campuses are growing these days thanks to the IoT which also promotes effective resource usage (Adam, Majid, & Ibrahim, 2023). By allowing students to experiment with their creativity around the campus, IoT enables interactive learning. Cloud computing and IoT devices that support university administration, instruction, and research are considered to be integrated into a smart campus (Alghamdi & Shetty, 2016; Min-Allah & Alrashed, 2020; Tiwari, Rosak-Szyrocka, & Żywiołek, 2022; Valks, Arkesteijn, Koutamanis, & Heijer, 2021). However, research by Alghamdi and Shetty (2016)) has divided IoT-based solutions into two groups, representing smart education ideas: smart campuses and smart universities. The authors therefore emphasized that the main focus of the smart university idea is on apps that would enhance the academic service administration and infrastructure of institutions. IoT services are developed by using sensors to turn ordinary things and solutions in the surroundings of educational institutions into intelligent objects or solutions (Cata, 2015; Roccotelli & Mangini, 2022). A common item is transformed into a complete intelligent gadget to facilitate intelligent decision-making in the campus setting (Jakubczyc & Owoc, 1998). In addition to supporting the learning cycle both inside and across the campus ecosystem, a smart campus facilitates the efficient creation of services and applications (Tsai & Yeh, 2015). This also enhances the quality of student graduation and boosts campus performance (Coccoli, Guercio, Maresca, & Stanganelli, 2014). Teaching, administration, and the service of campus stakeholders are the three facets of campuses that are made simpler by the implementation of the smart idea (Chuling, Xie, & Peng, 2009; Muhamad, Kurniawan, & Suhardi,, & Yazid, S., 2017). These are some IoT use cases in certain universities around the world that used the IoT on campus for better efficiency, sustainability, and a more conducive learning environment.

2.1.1 Arizona State University (ASU) – United States

IoT is enabling ASU to emerge as a smart campus. ASU uses an IoT-driven energy management system to monitor real-time energy consumption through sensors and smart devices on campus buildings. The smart lighting system reduces energy waste by automatically adjusting the lighting, heating, and cooling systems according to occupancy and environmental conditions. ASU's campus motion sensors and smart lighting also save energy and promote a greener environment. Besides curbing operational costs, AI and data analytics make the university resource-efficient by offering it predictive maintenance resources. The initiatives have brought increased energy efficiency and lower carbon emissions to ASU, supporting the university's sustainability goals, while improving comfort and operational efficiency on campus.

2.1.2 Green University: Vancouver Eco-friendly Campuses – Canada

Universities in Vancouver have adopted IoT in their quests to become sustainable and smart eco-friendly campuses. IoT-based energy management systems that monitor and control energy consumption at different campus facilities are one of the key initiatives. These systems use smart meters and sensors to offer detailed information on energy consumption, which enables universities to minimize waste and improve resource efficiency. For example, the Canadian University of British Columbia (UBC) has adopted an IoT-based campus management solution to get real-time readings of environmental parameters, including temperature, humidity, and air quality. The system encourages the occupants of buildings to only use energy on their terms by analysing and collecting data used to ascertain how green and comfortable the places we work and study are. As the campus cannot solely rely on a centralized monitoring of water use due to its size and complexity, IoT devices have been integrated into its water conservation systems so that the entire campus can reduce water use through its detection and irrigation management systems. The implementation of this new technology has put these institutions one step closer to reaching their sustainability objectives. In doing so, it not only ensured a lower carbon footprint but also highlighted one way other organizations might sustainably roll out smart city technologies (Ahmed et al., 2022; Majeed & Ali, 2018; Rabbi Shawon, Rahman Chowdhury, & Rahman, 2023). Table 2.1 shows the challenges and opportunities of IoT on campus.

The future of urban campus is one of the sub-systems of the larger, connected urban IoT ecosystem (Villegas-Ch, Palacios-Pacheco, & Luján-Mora, 2019; Zygiaris, 2013). As cities head in the direction of becoming increasingly smart, universities can be major players in the ecosystem. No longer self-contained institutions, IoT-connected universities will be embedded in a networked urban environment, both providing and receiving services to and from the wider smart city system (Al-Emran, Al-Sharafi, & Shaalan, 2023; Bandara et al., 2016). An example of this is the IoT systems of universities that

Table 2.1 Challenges and opportunities of IoT on campus

Challenges	*Opportunities*
Challenges in Security: Data breach risk, device vulnerabilities, and unauthorized access due to an unprotected and decentralized IoT system, which is further supplemented by weak authentication mechanisms.	**Smart Campus Operations**: Optimize the campus infrastructure using real-time data collection; build process automation systems such as building management system (BMS) and energy management system (EMS), throughout the integrated building and management systems.
Privacy Concerns: Due to the continuous collection of data from all around the campus by IoT devices, students and faculty are concerned about data privacy and surveillance.	**Sustainability**: Smart grid and IoT-based renewable energy systems can be integrated into campus infrastructure to improve IoT solutions to optimize energy consumption and mitigate waste.
Scalability and Complexity: As IoT hits the campus in a big way, the management of a large number of devices and applications across a decentralized campus is an issue with the integration of a large number of devices that are not supported on existing infrastructure.	**Innovative Solutions for Learning**: IoT contributes to classroom automation, instant monitoring, and interactive environments for learning that improve quality education.
System Management: How do different systems manage this – especially credentialing, course certificates, and how do they manage the flow of administrative and transactional data?	**Smart Services**: The IoT devices can turn ordinary items (such as the door, lighting, and equipment) into intelligent devices that enable more convenient services (smart attendance, smart parking, resource automation, etc.).
Integration with Existing IT Systems: IoT applications must co-exist with the current campus IT structures that are often not designed to effectively manage IoT infrastructure.	**Informed Decision-Making**: IoT can help in the collection of data which is required to enhance campus operation, research assistance, and in making decisions regarding campus rules and resource management.
Trustworthiness: Because IoT devices are connected to delicate networks, users are hesitant to use them due to reliability and trust in their ability to function.	**The IoT can also be utilized to enhance security and safety** on college campuses using surveillance systems as well as smart lighting and alert systems, providing a safer environment on campus.

Sources: own study based on Min-Allah and Alrashed (2020), Polin, Yigitcanlar, Limb, and Washington (2023), and Sneesl et al. (2022).

can be directly interfaced with city-wide smart grid systems resulting in efficient energy sharing and allocation between campus and city-integrated infrastructure (Ahmed et al., 2022; Heijer, Magdaniel, & Flavia, 2018). By integrating academic research on smart mobility urban transport systems, universities could optimize student transit along with the general public. Communication layers built on the IoT could integrate campuses into the larger city, allowing real-time data about urban services, events, or environmental factors to be shared. They are not only becoming more connected and efficient, but also pushing towards the vision ofbuilding inclusive and prosperous cities with the help of urban IoT systems. (Lin, Chen, Shieh, Lin, & Yen, 2018; Zanella, Bui, Castellani, Vangelista, & Zorzi, 2014). By driving research, building IoT solutions, and training the next generation of workers, universities are not only preparing their own campuses for the future but also cultivating the urban IoT systems that will work in the cities of tomorrow (Arenas et al., 2022; Schaffers et al., 2011; Villegas-Ch, Palacios-Pacheco, & Román-Cañizares, 2020).

2.2 Artificial Intelligence (AI) in Education and Administration

In 1956, John McCarthy came up with the phrase artificial intelligence (AI; Haenlein & Kaplan, 2019). AI, in general, is the development of machines that exhibit some level of intelligence and are able to experience cognition, learning, decision-making, and adaptation to the environment (Chen, Chen, & Lin, 2020). The use of AI has spread across departments inside educational institutions and other parts of the education industry. AI in education may significantly change how academic and administrative personnel use their time and how each student is treated on an individual basis (Li, Gobert, & Dickler, 2019). A number of significant effects have resulted from the use of AI in education, including increased effectiveness and efficiency in school administration, global learning, personalized and tailored learning, and smarter material (Timms, 2016). By giving teachers and students more authority, AI may enhance student learning (Popenici & Kerr, 2017). Tutors who provide individualized teaching and adaptive assessment, as well as classroom management systems driven by AI, have the potential to transform educational activities ('How "Intelligent" Tutors Could Transform Teaching', 2017). According to a poll, 44% of instructors said AI might assist with time-consuming, repetitive chores like making copies and collecting attendance ('Teachers, the robots are coming. But that's not a bad thing', 2020). According to a recent study, AI might free up almost one-third of instructors' workdays (Bryant, Heitz, Sanghvi, & Wagle, 2020) for more human-driven activities like mentoring students one-on-one and working with colleagues. There are two primary ways that AI applications are helping enterprises in the education industry (Chassignol, Khoroshavin, Klimova, & Bilyatdinova, 2018; Li & Jeong, 2020):

- academic level (such as evaluation, comments, tutoring, etc.),
- administrative level (such as library services, counselling, and admissions).

Figure 2.1 presents the most important benefits of AI in education and administration.

AI-driven personalized learning is one of the key contributions of AI to education (Ayeni, Hamad, Chisom, Osawaru, & Adewusi, 2024; Kaswan,

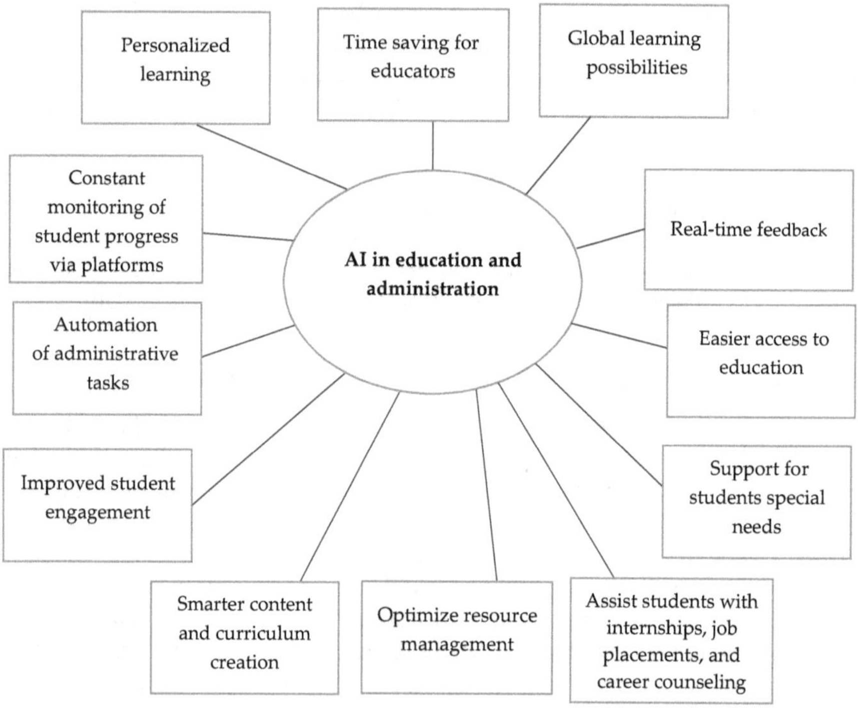

Figure 2.1 Benefits of AI in education and administration.

Source: own study.

Dhatterwal, & Ojha, 2024). AI systems employ algorithms to mine endless amounts of data on students, producing personalized learning paths for students and modifying elements such as content, pace, and assessments to suit specific needs. It is argued that a standard education does not suffice for every student, which is the reason personalized learning aims to improve the engagement and performance of learners (Kaswan, Dhatterwal, & Ojha, 2024; Shoaib et al., 2024). Instead, AI could potentially serve as a means for students to reach their learning goals through personalized adaptative learning, for example, intelligent tutoring systems (Dziuban et al., 2018), teacherbots (T-bots) (Bayne, 2015; Pillai, Sivathanu, Metri, & Kaushik, 2024), AI teaching assistants or tailored learning (Gillani et al., 2023). Experimental research in India finds that individualized technology-assisted after-school education can improve math and language test results (Muralidharan, Singh, & Ganimian, 2019). More recent AI technologies such as GPT-4 might lead to larger gains that are due to the tutoring effect. The use of generative AI has the possibility of enabling students as well as aiding them in enriching their learning experiences and resources (Dai, Liu, & Lim, 2023). Generative AI could play as a collaborator, coach, or tutor in the classroom in multiple ways (Mollick & Mollick, 2023).

As an alternative, AI may be used outside of the classroom to serve as a coach or tutor while knowledge-application activities take place in class (Katsamakas, Pavlov, & Saklad, 2024). AI-driven platforms can monitor a student's progress, identify strengths and gaps, and customize learning materials to respond. Adaptive learning, with some platform examples like Knewton and Smart Sparrow, in which machine learning algorithms provide feedback or recommendations to students so that they may continue their learning at their own pace (Yunusaliyev, Yuldashev, & Yusupov, 2024). Such systems also support educators through differentiated instruction. Several platforms and applications of AI are described by Chassignol et al. (Chassignol et al., 2018). These include interactive learning environments (ILEs), which are used to manage performance and provide feedback and exchanges between teachers and students; intelligent tutoring systems, such as MATHia, Why2Atlas, Comet, Viper, and ACTIVE Math, which have been used by educators or instructors for various subjects at different levels of education; and extensive use in learning assessment to track performance and improve the available pedagogical tools (Chassignol, Khoroshavin, Klimova, & Bilyatdinova, 2018).

The power of AI allows teachers to teach classrooms comprising students with different needs; it helps address the diverse needs of students, and especially those who require guidance (slow learners) or are ahead in their studies (fast learners) (Zou et al., 2024; 'Impact of AI on the inclusion of Learners with Special needs: Public Policy Perspective in Contemporary Scenario', 2024). It is important so that personalized learning can deliver more than adaptive and customized content delivery, it can also include real-time feedback and tracking of strengths and weaknesses (Contrino, Reyes-Millán, Vázquez-Villegas, & Membrillo-Hernández, 2024; Sajja, Sermet, Cikmaz, Cwiertny, & Demir, 2024). Due to the ability to differentiate instruction, teachers will be able to manage more students in the classroom with ease. In contrast to a traditional classroom setting, smart information may be accessed remotely, individually, and several times at once, making access to education easier, for example, video lectures and e-books. Students learn more and instructors are able to impart information more easily. AI can evaluate assignments, quizzes, and assessments within moments so that students can measure their progress immediately (Fagbohun, Iduwe, Abdullahi, Ifaturoti, & Nwanna, 2024; Halkiopoulos & Gkintoni, 2024). This not only saves precious time for students and also the instructors but also drives a more engaging experience as well. Automated feedback saves the instructor a tremendous amount of engagement time that can now be spent on higher levels of teaching where they're truly needed: for example, coaching critical thinking and giving personalized mentorship.

Thanks to AI, teachers are more productive and efficient, while students get more individualized and comprehensive instruction (Chen, Jiang, Yang, Yang, & Man, 2017; Fang et al., 2019). Chassignol et al. (Chassignol et al., 2018) emphasized the wide-ranging use of AI in subjects such as student evaluation, instructional strategies, content creation, and teacher–student communication. AI in education, according to Pokrivcakova, manifests as intelligent systems with adaptive capacities (Pokrivcakova, 2019). These principles and features of the systems allow AI in education to enhance students' learning experiences by coaching them and tailoring instruction to their needs and expectations, while also performing a variety of tasks that are typically or traditionally completed by instructors (Pokrivcakova, 2019). In their paper, Mikropoulos and Natsis also detail another facet of AI in education: virtual reality (VR) and three-dimensional (3-D) technology (Mikropoulos & Natsis, 2011). They note that VR offers a wealth of possibilities for the learning process, integrating 3-D technology with simulation since it allows for simulation and gives students the chance to learn through experience (Mikropoulos & Natsis, 2011). Education management, teaching, and learning may undergo a digital revolution with the use of AI and other Industry 4.0 technologies, such as the IoT (Quy, Thanh, Chehri, Linh, & Tuan, 2023; Rosak-Szyrocka, Żywiołek, Nayyar, & Naved, 2023; Tripathi & Rosak-Szyrocka, 2024), cobots (collaborative robots) that support instructors in the classroom (Rosak-Szyrocka, 2024; Timms, 2016), learning analytics, educational data mining, and intelligent web-based education (Chen et al., 2020). Author Timms (Timms, 2016) proposed that cobots – robots that collaborate with instructors or colleagues – are being used to teach kids basic skills like spelling and pronunciation while adapting to their individual capacities. In a similar vein, web-based and online education has evolved from merely providing students with online resources to download, study, and complete assignments in order to pass to incorporating intelligent and adaptive web-based systems that can recognize student and instructor behaviour and make necessary adjustments to enhance the learning environment (Chassignol et al., 2018; Peredo, Canales, Menchaca, & Peredo, 2011; Verstaevel, Boes, & Gleizes, 2017). Mobile education is advanced by the technological advancement of AI in mobile devices, which offers students convenience by assisting them in less time and achieving individualized and interactive learning. For example, since AI can link students to the virtual classroom, virtual reality makes it easier to build a global classroom and extend the learning process outside the physical classroom. Additionally, chatbots powered by AI provide individualized online instruction and transform teachers into chat partners. The degree of comprehension of the pupils may be evaluated by this technique (Chen et al., 2020).

AI is reshaping research and innovation at those same educational institutions (Abulibdeh, Zaidan, & Abulibdeh, 2024; Rosak-Szyrocka & Knop, 2024). The research process traditionally has been labour intensive – collecting, analysing, and synthesizing data have all been manual processes. AI is optimizing these processes by automating data analysis and allowing voracious collaboration between departments and outside institutions (Munagandla, Dandyala, & Vadde, 2024). Armed with machine learning algorithms and natural language processing tools, researchers are able to analyse massive datasets, discern the stories within the data, and glean insights in a fraction of the time that traditional methods would permit. AI can even identify gaps in the literature and recommend new avenues of research, in addition to its role in helping process scientific data and literature reviews. AI-powered applications help us predict results, or recommend paths, based on previously existing research, making it easier to get new findings for generating knowledge and drive innovation (Nayyar & Tripathi, 2024; Tripathi & Rosak-Szyrocka, 2024).

Besides revolutionising research, AI is improving the administrative aspects of universities as well (Hakimova Muhabbat et al., 2024; Villegas-Ch, García-Ortiz, Mullo-Ca, Sánchez-Viteri, & Roman-Cañizares, 2021). The use of AI in educational institutions takes many forms, such as the automation of administrative duties and procedures, curriculum and content creation, teaching, and the learning processes of students. According to Frank et al. (2019)), AI would change the way labour is distributed across sectors and vocations and reshape employment in the future. Over one-third of the careers that US students want are thought to be very vulnerable to automation over the next 15 years (Nakamura & Zeira, 2018). According to Herold ('Forty Percent of Elementary School Teachers' Work Could Be Automated By 2030, McKinsey Global Institute Predicts', 2019), AI is predicted to automate 40% of educators' current workloads over the course of the next 10 years, particularly non-teaching duties like monitoring student progress. By automating processes using web-based platforms or computer programs, AI has increased productivity in the execution of administrative duties including grading, examining student work, and giving feedback on assignments. Curriculum and content development, as well as instruction utilizing technologies like virtual reality, web-based platforms, robotics, video conferencing, audio-visual files, and 3-D technology, are additional educational domains where AI has been used to improve student learning. Administrative task automation, including scheduling, grading, and admissions, has improved system efficiency and alleviated staff burden dramatically. AI tools can automate tasks like grading assignments and giving feedback on student work. From these algorithms, these systems make use of machine analysis to provide an objective review of student performance and so the evaluation of student submissions. Automation of these tasks assists universities save time and resources and redirect administrative manpower to higher-priority activities; student engagement and resource management, for instance. AI is also an important player in the admissions process, as it can help facilitate the review of applications, respond to basic inquiries, and help potential students navigate the application process using chatbots and automated systems. For instance, the University of Texas Austin's (UTCS) Department of Computer Science unveiled GRADE (Graduate Admissions Evaluator), an explainable AI system that estimates the likelihood of a new applicant being accepted into their graduate program using logistic regression on historical admission data (Waters & Miikkulainen, 2014). The number of complete application evaluations and the amount of time experts spent reviewing each application were decreased by GRADE, even though it did not make the ultimate admission decision. To estimate an incoming applicant's prospective performance and select students best suited for the program (Zhao, Lackaye, Dy, & Brodley, 2020) analysed characteristics taken from students' application materials and their performance in the program of study. When it comes to student admissions, educational institutions consider yield rate – the percentage of admitted candidates who choose to enrol at a particular school – to be a crucial measure. Students' enrolment choices have been predicted by machine learning, which would assist the institution in strategically choosing students to increase yield rates and maximize resource allocation (Jamison, 2017). Furthermore, future success may be predicted by determining if students choose appropriate majors based on their individual backgrounds and past academic achievements. Classifying students into appropriate majors in an effort to position them for academic achievement has also been done using machine learning (Assiri, Bashraheel, & Alsuri, 2022). AI can help academic advisors mentor students, enhancing their learning experience, retention, and satisfaction. AI can also assist students with internships, job placements, and career counselling (Akiba & Fraboni, 2023). Many HEIs find that managing alumni relations is crucial, and AI facilitates this process (Daniel, 2015).

AI helps to reduce administrative workload and improving overall student experience in the admissions process – making it more efficient and responsive. AI also optimizes resource allocation and scheduling in addition to grading and admissions. AI-driven systems can generate ideal class schedules by factoring in instructor availability, room capacity, and student interest. Real-time data analytics from different school departments to predict potential scheduling conflicts and to resolve them automatically before they arise as it ensures better resource utilization and on-time access to proper classes. AI aids better utilization of university spaces like classrooms, libraries, student dorms, etc. AI systems can predict when resources will come under pressure and adjust schedules or assignments accordingly; they can optimize operational efficiency.

AI can be used by the IT department, for example, to automate processes and tasks and reduce the cost of maintaining the IT infrastructure. Data and AI can help all university departments become more efficient and reduce operating expenses. HEIs can use AI to transform into learning organizations that strive for continuous improvement while adjusting to environmental changes (Katsamakas et al., 2024). In addition to helping intellectually and administratively, AI applications (AIAs) are also increasing the efficacy of education. AIA assists educators with a range of responsibilities, including virtual reality (VR), learning analytics (LA), admissions, and grading/assessments (G/A). It frees up more time for teaching and student guidance by reducing the administrative duties of a teacher. AIA makes a substantial contribution to improving student learning, reducing teacher effort, efficiently and conveniently grading and assessing students, and assisting with several other administrative duties in the modern-day teaching profession. AI has substantially changed academic and administrative tasks in a variety of ways, including the admissions process, counselling, library services, evaluation, feedback, tutoring, and more (Ahmad, Alam, Rahmat, Mubarik, & Hyder, 2022). Any educational institution's admissions procedure benefits from the use of AI technologies and apps. The first step in the admissions process is submitting an online application to the admission department. In addition, the entrance form is accompanied with recurring questions from candidates and their parents (Goksel & Bozkurt, 2019), which might be challenging to respond to quickly. Education institutions are tackling the problem by using AI solutions, such as chatbots, to manage the influx of questions during the admissions process (Sneesl et al., 2022).

Although the advantages of employing AI in education and administration are evident, its use raises ethical concerns (Belk, 2021; Salloum, 2024; Stahl, 2021). The most pressing concern is the presence of bias in AI algorithms. These are foundational models: AI operates as a system of machines, learning autonomously from humans. It distributes knowledge, accepts input from its numerous specialized characteristics, and leaves the deliberate goal of bias elimination to the training data source. If historical data contains biases, an AI effectively (and quite sensibly) uses this information to draw conclusions and take appropriate action. When an AI system administers an admissions process, it relies on historically biased data that reflects societal socioeconomic disparities like eth nicity or sexual orientation. Consequently, it prioritizes the affluent over others, leading to discrimination. This is especially concerning in education, where equity is everything. To address its challenges, universities need to ensure that AI systems are developed using diverse, representative datasets. In addition, AI systems should undergo periodic audits to detect and ameliorate any biases that may surface. Universities need to make transparent the use of AI systems to make decisions, especially high-stake decisions, in areas like grading, admissions, and student evaluations. Another ethical issue is that of the transparency of AI decision-making (Cheong, 2024). In particular, the decisions that AI systems make about students' academic performance or opportunities need to be transparent to both students and faculty members. So, for example, if an AI system assesses a student's application for admission or assigns that student a grade, that student should have access to an explanation of the factors that shaped what the AI decided. This would mean creating systems that do not just allow students to understand what the system is saying but also to challenge or appeal decisions made by AI systems, where appropriate. This ensures fundamental alignment of AI as a transparent tool of human usage instead of a black box of responsibility where humans are in a position of zero accountability. In education, AI also raises a major concern of data privacy. AI systems typically need significant amounts of personal data, from students' academic records and performance data to behavioural data. This data needs to be kept safe from unauthorized access and misuse. Institutes and universities are very strict in terms of the data protection regulation, one of them that stands out is the General Data Protection Regulation (GDPR) in Europe and many other privacy laws around the world if not everywhere to ensure that students' data is protected. Moreover, students must retain ownership of their data and must also be able to offer consent for how it is processed through AI programs. For example, institutions should set forth clear boundaries regarding the collection and use of data as well as respect student privacy when it comes to data collection and analysis.

2.3 Cloud Computing and Digital Infrastructure

The rapid advancement of technology has caused a massive shift in the tertiary education industry in recent years. The indisputable significance of emerging technologies for education has prompted their application to higher education in colleges, universities, and institutes (Apostu, Vasile, Vasile, & Rosak-Szyrocka, 2022; Gupta & Rosak-Szyrocka, 2024; Rosak-Szyrocka, Apostu, & Pupp, 2024). Modern technology in higher education will upsurge academic outcomes and students' experiences (Grájeda, Burgos, Córdova, & Sanjinés, 2024; Quraishi, Ulusi, Muhid, Hakimi, & Olusi, 2024). Learning Management Systems (LMS) based on cloud computing is a great step forward in the administration and delivery of education (Banerjee, Kumar Mondal, Roy, Alnumay, & Biswas, 2024). This concept not only supports various aspects of online learning but also describes the utilization of cloud infrastructure for the storage, management, and dissemination of the educational materials (Sarosh, 2021). Everything about the learning process – resources, assignments, tests – can be shared and managed online, thus eliminating the need for local infrastructure. Cloud computing is one of the many innovations that have become the disruptive force capable to shift traditional teaching and learning paradigms (Guin et al., 2011; Guo & Zhang, 2015; Pan et al., 2023; Rosak-Szyrocka, 2024; Rosak-Szyrocka, Apostu, Ali Turi, & Tanveer, 2022).

Cloud computing is becoming more and more popular in higher education as it provides the on-demand integration of computer power (servers, storage, etc.) that is scalable. This paradigm change in IT infrastructure enables educational institutions to now enhance student and teacher outcomes, reduce the intensity of IT processes, and stimulate collaboration (Armbrust et al., 2010). Postsecondary education's greatest advantage of cloud computing is cost. Traditional IT infrastructures occasionally require large initial hardware and software purchases as well as continuous maintenance fees. But cloud computing reduces the significant expense of keeping large on-site infrastructure by allowing educational institutions to pay as they go (Balaji, Xu, Nwokafor, Gupta, & Agarwal, 2013; Mell & Grance, 2011). Using cloud-based tools, teachers and learners can access educational materials, applications, and collaboration platforms from anywhere electricity is accessible. This creates a more dynamic and inclusive learning space that is capable of meeting the diversification of many needs and interests of the student group in contemporary academic classes (Suhaib, 2020). With cloud-based collaboration tools such as Google Workspace and Microsoft 365, instructors and students can interact and collaborate more easily, creating more dynamic learning experiences and increasing student engagement. In the topic of cloud computing in the classroom, security and privacy have been raised as concerns. Therefore, sensitive student and research data needs to be properly protected, and all data privacy policy needs to be abided (Buyya, Yeo, Venugopal, Broberg, & Brandic, 2009). As a result, cloud service providers are addressing these issues due to the investment they made in strong security mechanisms and compliance frameworks, surpassing on-premises solutions in security (Subashini & Kavitha, 2011). Table 2.2. shows the benefits and challenges of cloud computing in education.

Table 2.2 Benefits and challenges of cloud computing in education

Benefits	*Challenges*
Cost Efficiency: Reduced capital expenditure on hardware and on-premises IT infrastructure, so that universities can instead direct financing resources to research, syllabus formation, and some other innovative projects.	**Data Security and Privacy Risks**: Universities typically store sensitive personal and academic data. Sensitive student records, research projects, and intellectual property may be vulnerable to data breaches or unauthorized access in cloud storage. Having proper encryption and compliance models is a necessity.
Scalability and Flexibility: Cloud computing allows universities to dynamically scale computing resources that fit peak periods (enrolment seasons or examination seasons). Agents retain the ability to optimize the resources they consume at the cost of having to forgo investments in permanent infrastructure.	**Vendor Lock-In**: Using specific cloud providers can restrict flexibility. There would also be operational risks; the adoption curve of migrating to alternative providers or towards hybrid solutions would require an investment in time and capital, as implementation and development would be costly to implement.
Global Accessibility: Cloud-based platforms enable students and faculty to access educational resources, tools, and systems from anywhere in the world, promoting global collaboration and remote learning opportunities for international students.	**Regulatory Compliance**: Universities are required to comply with a range of data protection laws, such as the General Data Protection Regulation (GDPR) in European Union nations or Family Educational Rights and Privacy Act (FERPA) in the United States. When the cloud providers for universities operate globally, that creates legal and regulatory complexities.
Collaborative Platforms: Cloud-based applications, such as Google Workspace or Microsoft Teams, support real-time collaboration on research, group projects, and administration, enabling team working between students, staff, and researchers. When the cloud providers for universities operate globally, that creates legal and regulatory complexities.	**Integration Challenges**: Legacy systems, such as on-premises databases and personalized software, do not integrate effortlessly with an existing cloud platform. Addressing these challenges can be resource-intensive in terms of both IT skills and infrastructure.
Enhanced Research Capabilities: Access to high-performance storage, computing, etc., as high-performance computers (HPC) needed to handle big data analytics, AI, genomics, environmental science, etc. Researchers are able to easily process and share vast datasets.	**Internet Dependency**: Cloud systems may not work properly in the absence of a stable and high-speed internet connection. Cloud adoption can impede institutional efficiency and delivery of education in resource constrained areas.

(*Continued*)

Table 2.2 (Continued)

Benefits	*Challenges*
Disaster Recovery and Backup: Automation and backup solutions for critical academic and administrative data are features of cloud systems that were not lost during outages caused by cyberattacks, natural disasters, or equipment failures.	**Technological Literacy Gaps**: Faculty, staff, and students may not have had prior exposure to cloud-based tools. This will require universities to implement training programs that lead to the effective and efficient use of these technologies.
Energy Efficiency: Universities can align with the sustainability goals through a decrease in energy consumption and carbon emissions in both IT resources, leading to a consolidation of IT resources and migratory steps towards the cloud infrastructure.	**Long-Term Costs**: Many cloud services provide subscription-based pricing models, which can lead to unexpected long-term costs. The high, and recurring, operational costs of running universities require a prudent and measured approach to budgeting.
Support for Multidisciplinary Research: The cloud allows for real-time collaboration between departments and institutions and provides the tools for data sharing and simultaneous access to large-scale datasets, thereby fostering interdisciplinary research.	**Trust and Dependency**: Heavy dependence on third-party vendors to provide essential services can compromise institutional control over data and operations, making them more dependent upon each other, especially during outages and service inadequacies.
Innovation in Education: The transformation of education through technology and innovation by the emergence ofpersonalized and adaptive learning experiences. Analytics enables personalized learning experiences by curating content based on students' unique requirements, thereby improving both engagement and outcomes.	**Cultural and Organizational Resistance**: A transition to cloud-based solutions may face resistance from staff or faculty, familiar with classical IT systems, leading to delayed adoption.

Sources: own study based on Cindy, Walilulu, Pattipeilohy, Mark, and Sudadi (2024) and Sarosh (2021).

References

Abulibdeh, A., Zaidan, E., & Abulibdeh, R. (2024). Navigating the confluence of artificial intelligence and education for sustainable development in the era of industry 4.0: Challenges, opportunities, and ethical dimensions. *Journal of Cleaner Production*, *437*, 140527, https://www.sciencedirect.com/science/article/pii/s0959652623046851

Adam, K., Majid, M. A., & Ibrahim, Y. (2023). Smart campus reliability based on the Internet of Things. In *Lecture notes in networks and systems*, edited by M. Al-Emran, M.A. Al-Sharafi, & K. Shaalan: *Vol. 550. International conference on information systems and intelligent applications* (pp. 353–360). Cham: Springer International Publishing.

Ahmad, S. F., Alam, M. M., Rahmat, M. K., Mubarik, M. S., & Hyder, S. I. (2022). Academic and administrative role of artificial intelligence in education. *Sustainability*, *14*(3), 1101.

Ahmed, M. A., Chavez, S. A., Eltamaly, A. M., Garces, H. O., Rojas, A. J., & Kim, Y.-C. (2022). Toward an intelligent campus: IoT platform for remote monitoring and control of smart buildings. *Sensors, 22*(23), 9045, https://www.mdpi.com/1424-8220/22/23/9045

Akiba, D., & Fraboni, M. C. (2023). AI-supported academic advising: Exploring ChatGPT's current state and future potential toward student empowerment. *Education Sciences, 13*(9), 885, https://www.mdpi.com/2227-7102/13/9/885?utm_source=flipboard&utm_content=dougballard568%2fmagazine%2fchatgpt%2f+ai+-%2f+gpt+4

Al-Emran, M., Al-Sharafi, M. A., & Shaalan, K. (Eds.) (2023). *Lecture notes in networks and systems: Vol. 550. International conference on information systems and intelligent applications*. Cham: Springer International Publishing.

Alghamdi, A., & Shetty, S. (2016). Survey toward a smart campus using the Internet of Things. In *2016 IEEE 4th international conference on future Internet of Things and Cloud (FiCloud)* (pp. 235–239). IEEE.

An, Q., Wang, H., & Chen, X. (2022). EPSDNet: Efficient campus parking space detection via convolutional neural networks and vehicle image recognition for intelligent human-computer interactions. *Sensors, 22*(24), 9835, https://www.mdpi.com/1424-8220/22/24/9835

Apostu, S. A., Vasile, V., Vasile, R., & Rosak-Szyrocka, J. (2022). Do smart cities represent the key to urban resilience? Rethinking urban resilience. *International Journal of Environmental Research and Public Health, 19*(22), 15410, https://www.mdpi.com/1660-4601/19/22/15410

Arenas, J., Palominos, P., Martin, R., Camargo, M., Bary, R., & Kaschel, H. (2022). Smart campus: Concepts, development and future. In *2022 IEEE 28th international conference on engineering, technology and innovation (ICE/ITMC) & 31st international association for management of technology (IAMOT) joint conference* (pp. 1–7). IEEE.

Armbrust, M., Fox, A., Griffith, R., Joseph, A. D., Katz, R., Konwinski, A., et al. (2010). A view of cloud computing. *Communications of the ACM, 53*(4), 50–58.

Assiri, B., Bashraheel, M., & Alsuri, A. (2022). Improve the accuracy of students admission at universities using machine learning techniques. In *2022 7th international conference on data science and machine learning applications (CDMA)* (pp. 127–132). IEEE.

Atzori, L., Iera, A., & Morabito, G. (2010). The Internet of Things: A survey. *Computer Networks, 54*(15), 2787–2805, https://www.sciencedirect.com/science/article/pii/s1389128610001568?casa_token=8wfuj9hhl-0aaaaa:eiprr5qox33vmd3xllo2gid5wo4dwsnhyh8c_cmqh8_mdlxfnu4hsv7joefsj0u7yrvdp7_rfo7p

Ayeni, O. O., Hamad, N. M. A., Chisom, O. N., Osawaru, B., & Adewusi, O. E. (2024). AI in education: A review of personalized learning and educational technology. *GSC Advanced Research and Reviews, 18*(2), 261–271, https://gsconlinepress.com/journals/gscarr/content/ai-education-review-personalized-learning-and-educational-technology

Balaji, B., Xu, J., Nwokafor, A., Gupta, R., & Agarwal, Y. (2013). Sentinel. In *Proceedings of the 11th ACM conference on embedded networked sensor systems*. New York: ACM.

Bandara, H. M. A. P. K., Jayalath, J. D. C., Rodrigo, A. R. S. P., Bandaranayake, A. U., Maraikar, Z., & Ragel, R. G. (2016). Smart campus phase one: Smart parking sensor network. In *2016 Manufacturing & Industrial Engineering Symposium (MIES)*. IEEE.

Banerjee, S., Kumar Mondal, M., Roy, M., Alnumay, W. S., & Biswas, U. (2024). A deep learning-based car accident detection framework using edge and cloud computing. *IEEE Access, 12*, 130107–130115.

Bayne, S. (2015). Teacherbot: Interventions in automated teaching. *Teaching in Higher Education*, *20*(4), 455–467.

Belk, R. (2021). Ethical issues in service robotics and artificial intelligence. *The Service Industries Journal*, *41*(13–14), 860–876.

Bhushan, B., Khamparia, A., Sagayam, K. M., Sharma, S. K., Ahad, M. A., & Debnath, N. C. (2020). Blockchain for smart cities: A review of architectures, integration trends and future research directions. *Sustainable Cities and Society*, *61*, 102360, https://www.sciencedirect.com/science/article/pii/s2210670720305813?casa_token=kgprwlept8maaaaa:jwkddiw2xlwpjged0nkoiwem3xui62ideecqymvbvz2a8c-nv00geuc_nkawn2_18w5ejyr2z02x

Bryant, J., Heitz, C., Sanghvi, S., & Wagle, D. (2020, January 14). How artificial intelligence will impact K–12 teachers. *McKinsey & Company*. Retrieved December 22, 2024, https://www.mckinsey.com/industries/social-sector/our-insights/how-artificial-intelligence-will-impact-k-12-teachers

Buyya, R., Yeo, C. S., Venugopal, S., Broberg, J., & Brandic, I. (2009). Cloud computing and emerging IT platforms: Vision, hype, and reality for delivering computing as the 5th utility. *Future Generation Computer Systems*, *25*(6), 599–616, https://www.sciencedirect.com/science/article/pii/s0167739x08001957?casa_token=6m6tljqzh-waaaaa:-8t_yh2smbzpfo05mbqwuyasnfgcdedem7gpevajrgm0lq2ebs-vb1-ywldgfcamywc1wrpjfrwv

Cata, M. (2015). Smart university, a new concept in the Internet of Things. In *2015 14th RoEduNet International Conference - Networking in Education and Research (RoEduNet NER)*. IEEE.

Chassignol, M., Khoroshavin, A., Klimova, A., & Bilyatdinova, A. (2018). Artificial Intelligence trends in education: A narrative overview. *Procedia Computer Science*, *136*, 16–24, https://www.sciencedirect.com/science/article/pii/s1877050918315382

Chen, H., Jiang, W., Yang, Y., Yang, Y., & Man, X. (2017). State of the art on food waste research: A bibliometrics study from 1997 to 2014. *Journal of Cleaner Production*, *140*, 840–846.

Chen, L., Chen, P., & Lin, Z. (2020). Artificial intelligence in education: A review. *IEEE Access*, *8*, 75264–75278.

Cheong, B. C. (2024). Transparency and accountability in AI systems: Safeguarding wellbeing in the age of algorithmic decision-making. *Frontiers in Human Dynamics*, *6*.

Chuling, L., Xie, Z., & Peng, P. (2009). A discussion on the framework of smarter campus. In *2009 Third international symposium on intelligent information technology application* (pp. 479–482). IEEE.

Cindy, A. H., Walilulu, H., Pattipeilohy, P., Mark, E., & Sudadi, S. (2024). Implementation of a cloud computing based learning management system in education management. *al-fikrah: Jurnal Manajemen Pendidikan*, *12*(1), 175.

Coccoli, M., Guercio, A., Maresca, P., & Stanganelli, L. (2014). Smarter universities: A vision for the fast changing digital era. *Journal of Visual Languages & Computing*, *25*(6), 1003–1011, https://www.sciencedirect.com/science/article/pii/s1045926x1400086x?casa_token=cllj6rexp48aaaaa:duswcycqbb607fvoyt2x3nkog1_ckxe2np_3wqqpnbf0ztbhdrnakpfrcwfpj2sqefpwympjify

Contrino, M. F., Reyes-Millán, M., Vázquez-Villegas, P., & Membrillo-Hernández, J. (2024). Using an adaptive learning tool to improve student performance and satisfaction in online and face-to-face education for a more personalized approach. *Smart Learning Environments*, *11*(1), 1–24, https://link.springer.com/article/10.1186/s40561-024-00292-y

Dai, Y., Liu, A., & Lim, C. P. (2023). Reconceptualizing ChatGPT and generative AI as a student-driven innovation in higher education. *Procedia CIRP*, *119*, 84–90, https://www.sciencedirect.com/science/article/pii/s2212827123004407

Daniel, B. (2015). Big Data and analytics in higher education: Opportunities and challenges. *British Journal of Educational Technology*, *46*(5), 904–920.

Dziuban, C., Moskal, P., Parker, L., Campbell, M., Howlin, C., & Johnson, C. (2018). Adaptive learning: A stabilizing influence across disciplines and universities. *OnlineLearning*, *22*(3), 7–39, https://eric.ed.gov/?id=ej1191489

Fagbohun, O., Iduwe, N. P., Abdullahi, M., Ifaturoti, A., & Nwanna, O. M. (2024). Beyond traditional assessment: Exploring the impact of large language models on grading practices. *Journal of Artificial Intelligence, Machine Learning and Data Science*, *2*(1), 1–8.

Fang, Y., Chen, P., Cai, G., Lau, F. C., Liew, S. C., & Han, G. (2019). Outage-limit-approaching channel coding for future wireless communications: Root-protograph low-density parity-check codes. *IEEE Vehicular Technology Magazine*, *14*(2), 85–93.

Forty percent of elementary school teachers' work could be automated by 2030, McKinsey Global Institute Predicts (2019, June 04). *Education Week*. Retrieved December 22, 2024, https://www.edweek.org/education/forty-percent-of-elementary-school-teachers-work-could-be-automated-by-2030-mckinsey-global-institute-predicts/2019/06

Frank, M. R., Autor, D., Bessen, J. E., Brynjolfsson, E., Cebrian, M., Deming, D. J., et al. (2019). Toward understanding the impact of artificial intelligence on labor. *Proceedings of the National Academy of Sciences of the United States of America*, *116*(14), 6531–6539.

Gillani, N., et al. (2023). Unpacking the 'Black Box' of AI in education. *Educational Technology & Society*, *26*(1), pp. 99–111. https://www.jstor.org/stable/48707970. Accessed 27 Mar 2025.

Goksel, N., & Bozkurt, A. (2019). Artificial intelligence in education: Current insights and future perspectives. In *Handbook of research on learning in the age of transhumanism* (pp. 224–236). IGI Global Scientific Publishing; IGI Global.

Grájeda, A., Burgos, J., Córdova, P., & Sanjinés, A. (2024). Assessing student-perceived impact of using artificial intelligence tools: Construction of a synthetic index of application in higher education. *Cogent Education*, *11*(1).

Gubbi, J., Buyya, R., Marusic, S., & Palaniswami, M. (2013). Internet of Things (IoT): A vision, architectural elements, and future directions. *Future Generation Computer Systems*, *29*(7), 1645–1660, https://www.sciencedirect.com/science/article/pii/s0167739x13000241?casa_token=qw03yxdkyveaaaaa:s2qrlbbb0xsfati9vt99fr3hmrf15of3fvpvplbozpe_frrhcuj-f3kcecldxsr5z28coudy0nxl

Guin, R. B., Chakrabarti, S., Tarafdar, C., Mandal, S., Banerjee, S., & Biswas, U. (2011). A smart architectural concept for making of a university education system using cloud computing paradigm. In *2011 World congress on information and communication technologies* (pp. 48–52). IEEE.

Guo, M., & Zhang, Y. (2015). "The research of smart campus based on Internet of Things & cloud computing," In *11th International conference on wireless communications, networking and mobile computing (WiCOM 2015), Shanghai*, pp. 1–6, doi:10.1049/cp.2015.0667

Gupta, S. K., & Rosak-Szyrocka, J. (2024). *Innovation in the University 4.0 System based on Smart Technologies*. CRC Press.

Haenlein, M., & Kaplan, A. (2019). A brief history of artificial intelligence: On the past, present, and future of artificial intelligence. *California Management Review*, *61*(4), 5–14.

Halkiopoulos, C., & Gkintoni, E. (2024). Leveraging AI in e-learning: Personalized learning and adaptive assessment through cognitive neuropsychology—a systematic analysis. *Electronics*, *13*(18), 3762, https://www.mdpi.com/2079-9292/13/18/3762

Haque, A. K. M. B., Bhushan, B., & Dhiman, G. (2022). Conceptualizing smart city applications: Requirements, architecture, security issues, and emerging trends. *Expert Systems*, *39*(5), e12753.

Heijer, A. C. den, & Magdaniel, C., Flavia, T. J. (2018). Campus–City relations: Past, present, and future. In *Knowledge and space* (pp. 439–459). Cham: Springer International Publishing.
Hossain, I., Das, D., & Rashed, M. G. (2019). "Internet of things based model for smart campus: challenges and limitations," In *2019 International conference on computer, communication, chemical, materials and electronic engineering (IC4ME2)*, Rajshahi, Bangladesh, pp. 1–4, doi:10.1109/IC4ME247184.2019.9036629
How 'Intelligent' tutors could transform teaching (2017, September 26). *Education Week*. Retrieved December 22, 2024, https://www.edweek.org/leadership/how-intelligent-tutors-could-transform-teaching/2017/09
Huang, Q., Rodriguez, K., Whetstone, N., & Habel, S. (2019). Rapid Internet of Things (IoT) prototype for accurate people counting towards energy efficient buildings. *Journal of Information Technology in Construction, 24*, 1–13.
Impact of AI on the inclusion of learners with special needs: Public policy perspective in contemporary scenario (2024).
Jakubczyc, J. A., & Owoc, M. L. (1998). Knowledge management and artificial intelligence. *Argumenta Oeconomica*, 1(6).
Jamison, J. (2017). Applying machine learning to predict davidson college's admissions yield. In *Proceedings of the 2017 ACM SIGCSE technical symposium on computer science education* (pp. 765–766). New York: ACM.
Kaswan, K. S., Dhatterwal, J. S., & Ojha, R. P. (2024). AI in personalized learning. In *Advances in technological innovations in higher education* (pp. 103–117). CRC Press.
Katsamakas, E., Pavlov, O. V., & Saklad, R. (2024). Artificial intelligence and the transformation of higher education institutions: A systems approach. *Sustainability, 16*(14), 6118, https://www.mdpi.com/2071-1050/16/14/6118
Kortuem, G., Bandara, A. K., Smith, N., Richards, M., & Petre, M. (2013). Educating the Internet-of-Things generation. *Computer*, *46*(2), 53–61.
Li, H., Gobert, J., & Dickler, R. (2019). *Evaluating the transfer of scaffolded inquiry: What Sticks and does it last?* (pp. 163–168). Cham: Springer.
Li, P., & Jeong, H. (2020). The social brain of language: Grounding second language learning in social interaction. *npj Science ofLearning*, *5*(1), 8, https://www.nature.com/articles/s41539-020-0068-7
Lin, Y.-B., Chen, L.-K., Shieh, M.-Z., Lin, Y.-W., & Yen, T.-H. (2018). CampusTalk: IoT devices and their interesting features on campus applications. *IEEE Access, 6*, 26036–26046.
Majeed, A., & Ali, M. (2018). How Internet-of-Things (IoT) making the university campuses smart? QA higher education (QAHE) perspective. In *2018 IEEE 8th annual computing and communication workshop and conference (CCWC)* (pp. 646–648). IEEE.
Malhotra, S., Kumar, A., & Dutta, R. (2021). Effect of integrating IoT courses at the freshman level on learning attitude and behaviour in the classroom. *Education and Information Technologies, 26*(3), 2607–2621, https://link.springer.com/article/10.1007/s10639-020-10376-0&casa_token=rua4shwad8saaaaa:vqv8-eo8xeol-u3upokaazikrbthyffbcvu4ee-dep0_xlqtbtwdqf7fksqyfffezwxvharhtvdwjlultjy
Mell, P. M., & Grance, T. (2011). *The NIST definition of cloud computing*. Gaithersburg, MD: National Institute of Standards and Technology.
Mikropoulos, T. A., & Natsis, A. (2011). Educational virtual environments: A ten-year review of empirical research (1999–2009). *Computers & Education, 56*(3), 769–780, https://www.sciencedirect.com/science/article/pii/s0360131510003052
Min-Allah, N., & Alrashed, S. (2020). Smart campus-A sketch. *Sustainable Cities and Society, 59*, 102231, https://www.sciencedirect.com/science/article/pii/s2210670720302183?casa_token=slbskdwyayyaaaaa:-sd-jvrujeeqqe3h8b0yxqrlxa0auogvojqrp3idojdrxluvkazal5i-_4qypjjwoznt1vu2r3qf

Mollick, E., & Mollick, L. (2023). *Assigning AI: Seven approaches for students with prompts*. Wharton School of the University of Pennsylvania & Wharton Interactive, June 11, 2023.

Monti, L., Tse, R., Tang, S.-K., Mirri, S., Delnevo, G., Maniezzo, V., & Salomoni, P. (2022). Edge-based transfer learning for classroom occupancy detection in a smart campus context. *Sensors*, *22*(10), 3692, https://www.mdpi.com/1424-8220/22/10/3692

Muhabbat, H., Mukhiddin, K., Jalil, H., Dustnazar, K., Farxod, T., Shavkat, M., et al. (2024). The digital frontier: AI-enabled transformations in higher education management. *Indonesian Journal of Educational Research and Technology*, *4*(1), 71–88, https://ejournal.upi.edu/index.php/ijert/article/view/60367/0

Muhamad, W., Kurniawan, N. B., Suhardi, & Yazid, S. (2017). Smart campus features, technologies, and applications: A systematic literature review. In *2017 International conference on information technology systems and innovation (ICITSI)* (pp. 384–391). IEEE.

Munagandla, V. B., Dandyala, S. S. V., Vadde, B. C. (2024). AI-driven optimization of research proposal systems in higher education. *Revista de Inteligencia Artificial en Medicina*, *15*(01), 650–672.

Muralidharan, K., Singh, A., & Ganimian, A. J. (2019). Disrupting education? Experimental evidence on technology-aided instruction in India. *American Economic Review*, *109*(4), 1426–1460, https://www.aeaweb.org/articles?id=10.1257/aer.20171112

Nakamura, H., & Zeira, J. (2018). Automation and unemployment: Help is on the way. *SSRN Electronic Journal*, June 26, 2018. Available at SSRN: https://ssrn.com/abstract=3202622 or http://dx.doi.org/10.2139/ssrn.3202622

Nayyar, S., & Tripathi, L. (2024). *Digital narratives in education*. OrangeBooks Publication.

Pan, M., Zhao, X., Lv, K., Rosak-Szyrocka, J., Mentel, G., & Truskolaski, T. (2023). Internet development and carbon emission-reduction in the era of digitalization: Where will resource-based cities go? *Resources Policy*, *81*, 103345, https://www.sciencedirect.com/science/article/pii/s0301420723000533

Peredo, R., Canales, A., Menchaca, A., & Peredo, I. (2011). Intelligent web-based education system for adaptive learning. *Expert Systems with Applications*, *38*(12), 14690–14702, https://www.sciencedirect.com/science/article/pii/s0957417411007962?casa_token=wki06gyexhuaaaaa:ey5xlgicl2qg1q6qoe9elfuicagmefadq4q1fcmyakfz7slhwbra9pg7_1gacndq6zyv7bfcnz9m

Pillai, R., Sivathanu, B., Metri, B., & Kaushik, N. (2024). Students' adoption of AI-based teacher-bots (T-bots) for learning in higher education. *Information Technology & People*, *37*(1), 328–355, https://www.emerald.com/insight/content/doi/10.1108/itp-02-2021-0152/full/pdf?casa_token=e8dgwsocricaaaaa:nru1rkoc3pw8k0qasqopn7erztkm6bc9p3xyomblgfymn0el--1tz7kzer_y0zfia8dad9kzxp2_lqzwmmb5edcsrty8dymhkbpjw6nmzs9hk3z2qa110g

Pokrivcakova, S. (2019). Preparing teachers for the application of AI-powered technologies in foreign language education. *Journal of Language and Cultural Education*, *7*(3), 135–153, https://sciendo.com/article/10.2478/jolace-2019-0025

Polin, K., Yigitcanlar, T., Limb, M., & Washington, T. (2023). The making of smart campus: A review and conceptual framework. *Buildings*, *13*(4), 891, https://www.mdpi.com/2075-5309/13/4/891

Popenici, S. A. D., & Kerr, S. (2017). Exploring the impact of artificial intelligence on teaching and learning in higher education. *Research and Practice in Technology Enhanced Learning*, *12*(1), 22, https://link.springer.com/article/10.1186/s41039-017-0062-8

Quraishi, T., Ulusi, H., Muhid, A., Hakimi, M., & Olusi, M. R. (2024). Empowering students through digital literacy: A case study of successful integration in a higher education curriculum. *Journal of Digital Learning and Distance Education*, *2*(8), 667–681, http://rjupublisher.com/ojs/index.php/jdlde/article/view/208

Quy, V. K., Thanh, B. T., Chehri, A., Linh, D. M., & Tuan, D. A. (2023). AI and digital transformation in higher education: Vision and approach of a specific university in Vietnam. *Sustainability*, *15*(14), 11093, https://www.mdpi.com/2071-1050/15/14/11093

Rabbi Shawon, R. E., Rahman Chowdhury, M. S., & Rahman, T. (2023). Transforming urban living in the USA: The role of IoT in developing smart cities. *International Journal of Machine Learning Research in Cybersecurity and Artificial Intelligence*, *15*(01). Available Online: https://ijmlrcai.com/index.php/Journal/index

Roccotelli, M., & Mangini, A. M. (2022). Advances on smart cities and smart buildings. *Applied Sciences*, *12*(2), 631, https://www.mdpi.com/2076-3417/12/2/631

Rosak-Szyrocka, J. (2024). The era of digitalization in education where do Universities 4.0 go? *Management Systems in Production Engineering*, *32*(1), 54–66, from https://sciendo.com/article/10.2478/mspe-2024-0006

Rosak-Szyrocka, J., Apostu, S. A., Ali Turi, J., & Tanveer, A. (2022). University 4.0 sustainable development in the way of Society 5.0. *Sustainability*, *14*(23), 16043, https://www.mdpi.com/2071-1050/14/23/16043

Rosak-Szyrocka, J., Apostu, S. A., & Pupp, Z. (2024). The relationship between technology of higher education in the case of European countries, a panel approach. *Polish Journal of Environmental Studies*, *33*(3), 2813–2821.

Rosak-Szyrocka, J., & Knop, K. (2024). Navigating the fourth industrial revolution: Insights from a comprehensive bibliometric study on Industry 4.0. *Production Engineering Archives*, *30*(4), 501–519.

Rosak-Szyrocka, J., Żywiołek, J., Nayyar, A., & Naved, M. (2023). *The role of sustainability and artificial intelligence in education improvement*. CRC Press.

Sajja, R., Sermet, Y., Cikmaz, M., Cwiertny, D., & Demir, I. (2024). Artificial intelligence-enabled intelligent assistant for personalized and adaptive learning in higher education. *Information*, *15*(10), 596, https://www.mdpi.com/2078-2489/15/10/596

Salloum, S. A. (2024). AI Perils in education: Exploring ethical concerns. *Artificial Intelligence in Education: The Power and Dangers of ChatGPT in the Classroom*, *144*, 669–675, https://link.springer.com/chapter/10.1007/978-3-031-52280-2_43

Sarosh, A. (2021). Machine learning based hybrid intrusion detection for virtualized infrastructures in cloud computing environments. *Journal of Physics: Conference Series*, *2089*(1), 12072.

Schaffers, H., Komninos, N., Pallot, M., Trousse, B., Nilsson, M., & Oliveira, A. (2011). Smart cities and the future internet: Towards cooperation frameworks for open innovation. In *Lecture notes in computer science* (pp. 431–446). Berlin, Heidelberg: Springer.

Shoaib, M., Sayed, N., Singh, J., Shafi, J., Khan, S., & Ali, F. (2024). AI student success predictor: Enhancing personalized learning in campus management systems. *Computers in Human Behavior*, *158*, 108301, https://www.sciencedirect.com/science/article/pii/s0747563224001699?casa_token=ld3vusqgrviaaaaa:d7pbh22lzfiklkpqft5kmndznejnxatzss4jtpidu1165hixxhztbf2dgm4qnulpdjykwipslciy

Sneesl, R., Jusoh, Y. Y., Jabar, M. A., & Abdullah, S. (2022). Revising technology adoption factors for IoT-based smart campuses: A systematic review. *Sustainability*, *14*(8), 4840, https://www.mdpi.com/2071-1050/14/8/4840

Stahl, B. C. (2021). Ethical Issues of AI. In *Artificial intelligence for a better future. SpringerBriefs in Research and Innovation Governance*. Cham: Springer. https://link.springer.com/chapter/10.1007/978-3-030-69978-9_4

Subashini, S., & Kavitha, V. (2011). A survey on security issues in service delivery models of cloud computing. *Journal of Network and Computer Applications*, *34*(1), 1–11, https://www.sciencedirect.com/science/article/pii/s1084804510001281?casa_token=dsatcvqmc7gaaaaa:gwucou__kgfivz2bq5w_mivddqmo3hixe2ekujoh72k-ln21wgdyqpdrkkwgiua5uyxcehv_xd5a

Suhaib, M. (2020). Usage of cloud computing technology and challenges in Japanese higher educational institutes. *International Journal of Scientific & Technology Research*, *9*(04), 2727–2733.

Sutjarittham, T., Habibi Gharakheili, H., Kanhere, S. S., & Sivaraman, V. (2018). Data-driven monitoring and optimization of classroom usage in a smart campus. In *2018 17th ACM/IEEE International Conference on Information Processing in Sensor Networks (IPSN)* (pp. 224–229). IEEE.

Teachers, the robots are coming. But that's not a bad thing (2020). https://www.edweek.org/technology/teachers-the-robots-are-coming-but-thats-not-a-bad-thing/2020/01

Timms, M. J. (2016). Letting artificial intelligence in education out of the box: Educational cobots and smart classrooms. *International Journal of Artificial Intelligence in Education*, *26*(2), 701–712, https://link.springer.com/article/10.1007/s40593-016-0095-y

Tiwari, S., Rosak-Szyrocka, J., & Żywiołek, J. (2022). Internet of Things as a sustainable energy management solution at tourism destinations in india. *Energies*, *15*(7), 2433, https://www.mdpi.com/1996-1073/15/7/2433

Tripathi, S., & Rosak-Szyrocka, J. (2024). *Impact of artificial intelligence on society*. Boca Raton: Chapman and Hall/CRC.

Tsai, I.-C., & Yeh, C.-H. (2015). Integrating SERVQUAL and importance-performance analysis for assessing smart campus service quality: A case study of an English training programme in Vietnam. In *2015 IEEE 15th international conference on advanced learning technologies* (pp. 436–440). IEEE.

Valks, B., Arkesteijn, M. H., Koutamanis, A., & Heijer, A. C. den (2021). Towards a smart campus: Supporting campus decisions with Internet of Things applications. *Building Research & Information*, *49*(1), 1–20.

Verstaevel, N., Boes, J., & Gleizes, M.-P. (2017). From smart campus to smart cities issues of the smart revolution. In *2017 IEEE smartworld, ubiquitous intelligence & computing, advanced & trusted computed, scalable computing & communications, cloud & big data computing, internet of people and smart city innovation (SmartWorld/SCALCOM/UIC/ATC/CBDCom/IOP/SCI)* (pp. 1–6). IEEE.

Villegas-Ch, W., García-Ortiz, J., Mullo-Ca, K., Sánchez-Viteri, S., & Roman-Cañizares, M. (2021). Implementation of a virtual assistant for the academic management of a university with the use of artificial intelligence. *Future Internet*, *13*(4), 97, https://www.mdpi.com/1999-5903/13/4/97

Villegas-Ch, W., Palacios-Pacheco, X., & Luján-Mora, S. (2019). Application of a smart city model to a traditional university campus with a big data architecture: A sustainable smart campus. *Sustainability*, *11*(10), 2857, https://www.mdpi.com/2071-1050/11/10/2857

Villegas-Ch, W., Palacios-Pacheco, X., & Román-Cañizares, M. (2020). An internet of things model for improving process management on university campus. *Future Internet*, *12*(10), 162, https://www.mdpi.com/1999-5903/12/10/162

Waters, A., & Miikkulainen, R. (2014). GRADE: Machine-learning support for graduate admissions. *AI Magazine*, *35*(1), 64–75, https://ojs.aaai.org/aimagazine/index.php/aimagazine/article/view/2504

Yunusaliyev, M., Yuldashev, I., & Yusupov, A. (2024). Revolutionizing the present education system through AI-driven learning platform with hybrid techniques. In *2024 4th international conference on advance computing and innovative technologies in engineering (ICACITE)* (pp. 523–527). IEEE.

Zanella, A., Bui, N., Castellani, A., Vangelista, L., & Zorzi, M. (2014). Internet of Things for smart cities. *IEEE Internet of Things Journal*, *1*(1), 22–32.

Zhao, Y., Lackaye, B., Dy, J. G., & Brodley, C. E. (2020). A quantitative machine learning approach to master students admission for professional institutions. *International Educational Data Mining Society*, https://eric.ed.gov/?id=ed607806

Zou, B., Liviero, S., Ma, Q., Zhang, W., Du, Y., & Xing, P. (2024). Exploring EFL learners' perceived promise and limitations of using an artificial intelligence speech evaluation system for speaking practice. *System*, *126*, 103497, https://www.sciencedirect.com/science/article/pii/s0346251x24002793

Zygiaris, S. (2013). Smart city reference model: Assisting planners to conceptualize the building of smart city innovation ecosystems. *Journal of the Knowledge Economy*, *4*(2), 217–231.

3 Sustainable Campus Development within Smart Cities

3.1 Green Building Practices and Smart Architecture

The term 'green campus' comes from the concepts of 'sustainable development' and 'environmental protection' of 1970s. The Talloires Declaration was signed by more than 350 nations in 1990, and 'environmental education' was integrated into the university curriculum (Ragazzi & Ghidini, 2017). Universities frequently feel obligated to address their own affects by implementing eco-friendly methods and 'greening' their campuses because of their role in increasing global temperatures (Global approaches to sustainability through learning and education, 2019; Smart education and learning market | Global Report [2027], 2024; *Technology management and its social impact on education*, 2023). The catch-all term for all these things is 'green-smart university', which includes things like green buildings, sustainable transportation, trash reduction, and energy efficiency. Using green-smart elements (waste reduction, energy efficiency, sustainable mobility, and improved resource management), universities seek to create a campus that is environmentally responsible as well as technologically competent (Pupiales-Chuquin, Tenesaca-Luna, & Mora-Arciniegas, 2022). Smart campuses, which are proficient at harnessing technology for optimizing the utilization of resources and the provision of services (Karvonen, Martin, & Evans, 2018; Majeed & Ali, 2018; Omotayo, Moghayedi, Awuzie, & Ajayi, 2021), hold an appeal that is more than just for improvements in efficiency. A growing body of evidence shows measurable benefits of these campuses, including reduced energy consumption, increased employee productivity, improved education, enhanced resource efficiency, and more community involvement. Digital transformation can significantly enhance university reputation, competitive advantage, and – most importantly – student engagement (Dai, Liu, & Lim, 2023; Huertas et al., 2021). According to the US Environmental Protection Agency, green building is 'the practice of creating structures and using processes that are environmentally responsible and resource-efficient throughout a building's life-cycle from siting to design, construction, operation, maintenance, renovation, and deconstruction' ('Basic Information | Green Building |US EPA', 2016; Makhloufi, Vasa, Rosak-Szyrocka, & Djermani, 2023; Shaheen et al., 2022). Green building is done to improve the quality of life of the people in the building

DOI: 10.4324/9781003621829-4

while at the same time reduce the negative social and environmental impacts of the built environment. Therefore, the measures involved in green building encourage more sustainable demand on energy management, water quality conservation management, waste management practices, air quality, site selection, building design, and materials to ensure good health and ecologically kind features (Oluwunmi, Oladayo, Role, & Afolabi, 2019). Table 3.1 compares the effects of green and traditional campuses on campus performance.

Table 3.1 Compares the effects of green and traditional campuses on campus performance

Metric	*Green campus*	*Traditional campus*
Energy Usage	Green campuses utilize renewable energy resources like solar panels, wind turbines, and geothermal systems. High-efficiency 'smart appliances' and networks monitor and optimize usage.	Traditional campuses rely heavily on grid electricity, much of which is from fossil fuel sources, with limited energy-efficient technologies, resulting in high consumption and wastage.
Water Consumption	Green campus has systems to reuse rainwater and greywater, low-flow fixtures to reduce usage, and reuse water for landscaping and non-potable uses.	Traditional campuses rely extensively on traditional water infrastructure systems with high wastage, with little integration of recycling or conservation tech.
Student Well-being	Optimized for natural lighting, air flow, and thermal comfort. Low-VOC materials are used in indoor spaces to be healthy and low stress. Greenery improves mental health and quality of life.	Many indoor spaces do not provide sufficient natural light and ventilation. Little consideration of health-promoting materials or environments could support work performance and comfort.
Carbon Footprint	Low emissions via renewable energy, sustainable materials, energy-efficient designs, and certifications like LEED or BREEAM. These lifecycle practices limit embodied and operational carbon.	High emissions because of dependence on traditional construction materials, fossil fuels, and inefficient building systems. No plans for measuring or lowering carbon footprint.
Operational Costs	The costs in the lifecycle are much lower for green technologies despite high upfront investments. Intelligent maintenance systems reduce time lost to repairs and prolong the life of infrastructure.	High operational costs as a result of inefficient energy use, outdated infrastructure, and increasing dependence on non-renewable resources.
Waste Management	Zero-waste goals drive campuses to recycle and compost most of their waste. Waste-to-energy programs generate electricity or biogas with organic waste, decreasing dependence on waste deposited in landfills.	Recycling and composting options – so few recycling and composting options. Waste that isn't recycled typically goes to landfills, where it can cause environmental harm and increase the cost of disposal.

(*Continued*)

Table 3.1 (Continued)

Metric	*Green campus*	*Traditional campus*
Campus Ecosystem	Encourages biodiversity with green roofs, vegetative open spaces, and sustainable landscaping. Reduces heat islands, equips with green designs, and enhances stormwater management.	Many traditional campuses did not explicitly prioritize the inclusion of green or biodiversity within their landscapes. Landscaping on such campuses often prioritizes aesthetics over ecological function, and many use non-native, high-maintenance, ornamental plants. Large quantities of water are also used for irrigation of these plants, especially in dry and arid climatic regions or dry seasons that make this practice inefficient in water utilization. The lack of enough greening in the final project also contributes to increase in temperature in concrete and traffic areas, referred to as the urban heat island effect. These campuses lack vegetation that could provide shade and cool the air through the process of evapotranspiration, thereby adding to localized rising temperatures. This not only increases the energy demands for cooling systems in buildings but also affects the comfort of students, faculty, and visitors outside.
Innovation and Leadership	Green campuses are living laboratories: sustainable practices are not only implemented but studied, tested, and improved upon with educators, researchers, and students collaborating to find solutions to environmental challenges. They are often held up as case studies for sustainable development and brand themselves as climate action leaders.	Traditional campuses are slow to embrace leading-edge sustainability practices, leaving themselves behind in terms of leadership opportunities and prominence.

Source: own study.

Table 3.2 shows case studies of universities that have successfully adopted green building practices.

Many higher education institutions throughout the world have adopted programs like seeking to become 'carbon-neutral universities' or changing into carbon-neutral organizations with minimal emissions. Universities want to be

Table 3.2 Case studies of universities that have adopted green building practices

University	*Country*	*Description of case study*
University of British Columbia (UBC)	Canada	UBC has been a leader in green building design for many years. One of the biggest achievements is the construction of one of North America's greenest buildings, the Centre for Interactive Research on Sustainability (CIRS). The CIRS building is powered by solar energy, uses natural light, and collects rainwater. The building is also net-positive for water and energy; it generates more energy than it consumes while harvesting rainwater for later use. UBC involves smart building management systems that monitor water, energy, and air quality in real time. Automatic controls reduce energy used even more by managing ventilation and lighting based on occupancy and outside conditions.
National University of Singapore (NUS)	Singapore	NUS has embraced green architecture with its net-zero energy School of Design and Environment (SDE) 4 building. NUS has cleverly architected its way around some of the challenges of Singapore's tropical environment. The SDE 4 building integrates solar panels, renewable materials, and passive design strategies (such as passive ventilation and shading). Due to the architecture's focus on natural ventilation, the hot and humid climate of Singapore needs less air conditioning.
Stanford University	United States	Stanford University's Jasper Ridge Biological Preserve Field Station and any of many other Stanford campus buildings are examples of the combination of smart technology and green construction techniques to create a sustainable academic venue. Stanford's green buildings incorporate energy-efficient HVAC systems, green roofs, and recycled materials. The Jasper Ridge building also employs geothermal heating and cooling, as well as solar power production, to reduce its carbon footprint.
Chulalongkorn University	Thailand	By incorporating smart design and green construction techniques, water and energy conservation and energy efficiency have been improved, rendering the property a more environmentally friendly urban campus in the heart of the city of Bangkok. In addition to this, the students have social and psychological benefits from green space including a better campus life and well-being.

(*Continued*)

Table 3.2 (Continued)

University	*Country*	*Description of case study*
The University of Sydney	Australia	Sustainable construction methods have been implemented extensively by The University of Sydney, resulting in better sustainability outcomes and significant reductions in operating costs. By incorporating systems for dynamic energy management utilizing smart technology, the university successfully monitors its energy consumption and continually seeks to improve efficiency. Such efforts align with the school's carbon reduction goals, asserting that next-generation technology and sustainable design can deliver both environmental and financial returns.
Technische Universität Berlin (TU Berlin)	Germany	TU Berlin has been a pioneer in the areas of smart buildings and sustainable design. The EUREF Campus is one well-known model for how a school might integrate sustainability in its flagship infrastructure. Energy-efficient buildings on the campus take advantage of passive cooling systems, solar energy, and geothermal energy.

Source: own study.

carbon neutral, but they also want to limit the consumption of materials and resources, enhance the quality of the environment, expand green spaces, and employ green transportation. In their efforts to promote sustainable development, universities also see campus greening initiatives and climate change education as essential elements of their plans (Kelly et al., 2022). Using green techniques at universities shows that they care about the environment and understand their social duty (Momani, Nour, & Jamaludin, 2020). The idea of 'Zero-Carbon Schools and Campuses' is a radical approach to education that places an emphasis on strong community participation, environmental responsibility, and adaptation in the face of climate-related concerns. Educational institutions are seen as more than merely academic settings in this holistic perspective. In order to reduce carbon emissions, increase resilience to natural catastrophes, and build strong relationships with their local communities, they are envisioned as living laboratories for sustainability, where professors, staff, and students actively contribute (Aghamolaei & Fallahpour, 2023; Moghayedi, Michell, Hübner, Le Jeune, & Massyn, 2024).

Aiming to improve education, increase awareness, and build institutional and human capacity in areas including early warning systems, impact reduction, and climate change mitigation, universities work to achieve SDG 13, particularly Target 13.3. Thomashow emphasizes how local actions may have a worldwide influence by tying the campus to the environment (Thomashow, 2014). Figure 3.1 illustrates a list of nine components for a sustainable campus, which are intended to inspire a 21st-century catalogue of transformative sustainable activities.

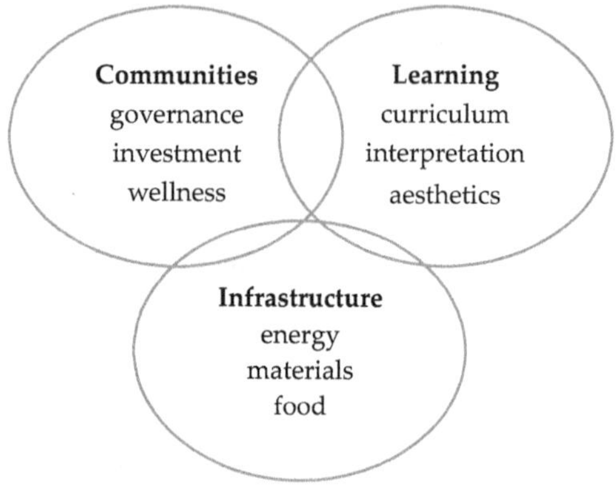

Figure 3.1 The nine elements of a sustainable campus.
Source: own study based on Thomashow (2014).

Research, teaching, student initiatives, community service, campus operations, housing, transportation, and food services are all ways they might contribute in a special way. University participation in these activities may significantly contribute to the reduction of greenhouse gas emissions, which is crucial for the global environmental agenda (Thakur, 2007).

3.2 Renewable Energy and Resource Management

Students on today's campuses are helping to lead the way now, and will soon be the leaders in business and government who will be called on to address this ongoing worldwide threat.
—Al Gore, former Vice President, Broadcast of NWF Chill Out: Campus Solutions to Global Warming

Given their significance to society, universities' sustainability is especially crucial. Nonetheless, the majority of schools are now dealing with issues including excessive resource and energy usage as well as serious environmental harm (Gholami et al., 2020). In China, for instance, the average energy and water consumption per student is significantly higher than the average per capita level. Primary and secondary schools use approximately 15% of all urban domestic water consumption, while the energy consumption of campuses makes up 5% of all national building energy consumption, a percentage that is steadily rising (Teah, Yang, Onuki, & Teah, 2019). Worldwide, campuses are moving to use sustainable energy sources like geothermal systems, solar panels, and wind turbines (Weijermars et al., 2018). They play a part in global perforations towards combating climate actions to minimize global warming problems ('Global Approaches to Sustainability Through Learning and Education').

The area of energy consumption is engaged in all stages of campus projects, including planning, designing, building, and operating, and it significantly affects the long-term viability of green campuses. Different levels of energy-saving design are needed during the planning and design stage, depending on the applicable laws and regulations of various regions. For instance, in some areas, public buildings must have a comprehensive energy-saving index of 65%. Within the same regions, the green building design scheme also takes into account the design of renewable energy sources like solar and air energy, water-saving equipment, and energy-saving lamps and lanterns. The green campus operation dynamically reflects all three of the energy-saving building, equipment, and renewable energy-use indices (Al-Dmour, 2023; Kanelli, Vardaka, Malesios, Katima, & Kalantzi, 2024; Li, Chen, & Yu, 2024). The primary goal of the material use is to thoroughly assess the longevity, safety, and environmental protection of the campus activity and decorating materials utilized in the campus's sustainable growth (Dong, Zhang, Yip, Swift, & Beswick, 2020).

Geothermal technologies, such as earth-integrated structures, earth-sheltered buildings, and geothermal ground-source heat pumps (GHPs), are being used on an increasing number of campuses. Lake cooling, direct-use high-temperature geothermal, and electricity-generating geothermal are a few additional examples. Rising energy prices and growing concerns about climate change are the main causes of the increase in the number of systems. For buildings that need heating, ventilation, and air conditioning (HVAC) systems, geothermal resources provide an effective, sustainable substitute (Cross, Eagan, Tolmé, Keniry, & John, 2011). Geothermal energy and other carbon-neutral renewables may provide large, sustained savings. For example, Ball State University in Indiana (United States) has already begun construction on a large-scale campus-wide geothermal system to provide heating and cooling energy for the whole school after realizing the potential monetary benefits (Dowling, Neumann, & Florea, 2016; Estep, n.d.). Additionally, by putting significant parts of a number of buildings underground, the University of Minnesota (United States) has reduced its energy use (Carlson et al., 2022). Because they have a basement, universities are able to profit from the thermal advantages of being below grade. For example, earth-bermed structures employ dirt built up along external walls and even on roofs to save energy expenses and protect the building from sun, wind, storms, and rain. If a structure is partially or completely below grade, the advantages of moving underground are highest. Subterranean structures' roofs may be at ground level, but they often have cleverly planned entrances, sunken courtyards, and banks of windows and skylights to bring in natural light and fresh air. Universities typically choose to move underground due to space constraints on campus. Because subterranean structures use the earth's constant temperatures and insulating qualities, they use about half as much energy to heat and cool as traditional buildings, which drastically reduces greenhouse gas emissions (Li, Zhao, Xie, Xiao, & Chuai, 2025; Toutouh & Alba, 2022).

Since the cost of installing photovoltaic (PV) solar panels has decreased dramatically over the past ten years and since solar usage is being increased by government incentives such as those included in the Inflation Reduction Act, more higher education institutions are thinking about doing so. PV panels may be retrofitted, installed as standalone solar arrays, or integrated into new construction to provide value to roofs, parking garages, parking lots, and other areas on university and college campuses. This energy source may not only lower carbon emissions but also operate as an example for students who want to learn about clean energy techniques that are sustainable. For instance, a tiny college in Connecticut recently celebrated the installation of a 770-kilowatt photovoltaic solar power system in its carports, which are spread over the campus's parking lots. It is anticipated that the project would compensate for around 30% of the school's yearly power use. Renewable energy is being adopted by educational institutions as we transition to a decarbonized future. Many American educational institutions now completely rely on renewable energy sources for their energy needs. Table 3.3 shows case studies of chosen TOP solar universities in 2022 in the United States.

University campuses may play a significant role in spearheading the shift to entirely clean, renewable energy by installing wind energy systems. Although it often encounters challenges, wind energy is safe, effective, limitless, and almost completely pollution-free. University campuses are particularly well-suited to wind power ('Wind Resource Data, Tools, and Maps | Geospatial Data Science | NREL', 2025; Apelfröjd, Eriksson, & Bernhoff, 2016):

Table 3.3 Case studies of chosen TOP solar universities in 2022 in the United States

University	*Description of case study*
Arizona State University – Arizona	The installation of 90 solar panels on four campuses allowed Arizona State University (ASU) to achieve its greenhouse gas emission target. ASU has installed photovoltaic solar panels, solar collectors, solar-shaded parking spots, and solar-shaded stadium seats as part of its on-campus solar systems. Together, these solar systems provide 24.1 MW of power, which covers half of the university's peak daily electricity needs. Over the course of a year, this helps ASU prevent CO_2 emissions equal to 4,804 passenger cars. An additional 28.8 MW of solar power is produced via off-site solar systems that the institution maintains in collaboration with Arizona Public Service Utility Company.
Stanford University – California	With 82% of its electricity coming from solar power, Stanford has installed both on-site and off-site solar systems as part of its goal to use only renewable energy by the end of 2022. With one off-site solar farm already in operation and another in planning, Stanford's on-site photovoltaic systems produce 4.9 MW of electricity, which is more than adequate to meet the university's energy needs. The present off-site solar farm does not provide Stanford with power directly; instead, the energy produced is sent to the California grid, which subsequently receives grid electricity equal to the solar farm's energy production.

(*Continued*)

Table 3.3 (Continued)

University	*Description of case study*
Rutgers University – New Jersey	Rutgers University is working towards achieving its 2030 goal of having zero carbon emissions from power purchases by putting solar solutions into place. The 32-acre solar canopy atop a parking lot is one of the biggest installations of its sort on a campus in the United States. It produces 8.8 MW of electricity, which is sufficient to meet 60% of the Rutgers Livingston campus's energy requirements. Through the sale of Solar Renewable Energy Credits (SRECs) and a lower power bill, Rutgers is expected to save $28 million over the course of 20 years from this 32-acre solar canopy. In addition to the parking lot cover, a 7,993-panel solar farm supplies an extra 10% of the campus's energy requirements. By producing 1.4 MW of solar electricity, the 7-acre solar farm lowers the university's yearly CO_2 emissions by 1,300 tons.
University of Arizona – Arizona	A centre for testing, assessing, and showcasing solar energy technology is the University of Arizona's Solar Zone. Additionally, this Solar Zone produces 25 MW of power. On a campus rooftop, the institution also has an agrivoltaic project. The 288 solar panels that make up this agrivoltaic scheme work in tandem with agricultural output to provide 15% of the building's power. Agriculture and solar panels work together to their mutual advantage because the plants cool the panels, increasing their efficiency, while the panels shield the plants from the sun.

Source: own study based on Wlda (2022).

- Location: A lot of colleges can set up micro-turbines on roofs or large turbines on broad fields, thanks to their ideal wind installation sites.
- Cost-effective: In many cases, onshore wind energy is less expensive than fossil fuel energy, particularly when tax incentives are taken into consideration.
- Educate and Research Opportunities: On-campus wind energy also offers engineering schools the chance to conduct innovative research and educate future professionals in the renewable energy sector.
- Finances: Universities may also build wind facilities on campus without incurring upfront capital expenditures by entering into power purchase agreements with utilities.
- Variations in Energy Output: Universities are coming up with ways to cope with different wind speeds. Case Western Reserve University, for example, views its campus as a 'living laboratory' and employs the Department of Energy's VOLTTRON software to reduce the variable output from its wind turbines.

Table 3.4 shows case studies on universities leading in renewable energy and resources management.

Table 3.4 Case studies of universities leading in renewable energy and resources management

University	*Country*	*Description of case study*
University of California, Davis (UC Davis)	United States	UC Davis is known worldwide for its innovations in renewable energy, including a 60-acre solar farm that provides 14% of the university's total electricity needs. The utilization of this renewable source is integrated into the smart grid of the campus, which allows real-time monitoring of and response to the use of energy. The university also uses geothermal systems for the efficient heating and cooling of campus buildings. These measures have dramatically decreased the university's carbon footprint and operating costs while providing a template for integrating large-scale renewable energy in higher education.
University of New South Wales (UNSW)	Australia	The University of New South Wales (UNSW) will be the first in the world to go 100% solar powered through sourcing energy from a large off-campus solar farm. This partnership enables the university to power all of its energy consumption with clean, renewable electricity. Smart energy management systems monitor usage, covering consumption across campus to maximize efficiency and minimize waste. This initiative is a major step to achieving the university's goal of considering 100% greenhouse gas reductions, and it shows that large institutions can be powered entirely by solar energy.
Stanford University	United States	Stanford established an innovative SESI program with a central energy facility fuelled by solar panels and geothermal energy. The system provides electricity, heating, and cooling to most of the buildings on campus while lowering greenhouse gas emissions by 68% and reducing water consumption by 18%. Smart energy dashboard – monitoring of energy consumption, allowing for efficient resource management and optimization of energy use in urban systems
University of British Columbia (UBC)	Canada	UBC generates heat for its campus with a biomass energy system that greatly reduces carbon emissions. Additionally, campus integrated solar photovoltaic systems and a district energy system, which real-time monitor renewable and traditional energy sources for optimized end use. Such initiatives have made UBC one of the most energy-efficient campuses in North America, conserving both operating costs and environmental resources.

(*Continued*)

Table 3.4 (Continued)

University	*Country*	*Description of case study*
Nanyang Technological University (NTU)	Singapore	NTU adopts renewable energy at many levels, including solar-powered buildings and innovative energy management systems that reduce the electricity consumption – by 30% no less. The utilization of Singapore's scarce wind resources for energy generation led us to establish a wind turbine research program on campus as well. Smart energy analytics and IoT sensors are used throughout the campus to constantly gauge and optimize energy utilization.
ETH Zurich,	Switzerland	ETH Zurich powers its campus with a hybrid energy system consisting of geothermal heat pumps, solar panels, and hydropower. To be sustainable not only by reducing fossil fuel dependence but also by operating all year round. They also leverage cutting-edge monitoring technologies to process energy efficiency data, ensuring that the campus is constantly optimizing operations through data-derived strategies.

Source: own study.

3.3 Waste Management and Water Conservation Technologies

One of the crucial elements of the green-smart institution is waste management (Chen, Jiang, Yang, Yang, & Man, 2017; Omotayo, Olanipekun, Obi, & Boateng, 2020). Paper, plastic, and food waste are among the many waste products produced by universities. Initiatives to reduce waste seek to increase the quantity of material that is recycled or reused while reducing the amount of garbage that is produced (Ardito, Ferraris, Messeni Petruzzelli, Bresciani, & Del Giudice, 2019; Dahlawi & El Sharkawy, 2021; Shankar & Khandelwal, 2017). These projects include the installation of composting facilities, the distribution of recycling bins around campus, and the digitization of paper products to cut down on waste ('Smart University: A University In the Technological Age', 2021; Jakimiuk et al., 2023). For instance, more than 70% of Cornell University's garbage is recycled or composted. Similarly, according to Adeniran, Nubi, and Adelopo (2017)) and Gursoy Haksevenler, Kavak, and Akpinar (2022)), the University of Florida recycles 30% of its garbage, whereas the University of Lagos recycles 53% of its garbage by using integrated waste management techniques.

Another popular approach for improving waste management is waste diversion, which involves lowering the amount of garbage that is sent for disposal (i.e., landfilling, incineration, or even littering). Waste reduction and zero waste (ZW) are related concepts. By reducing trash creation, reusing items, and preparing garbage for recycling – which necessitates separation at the source – both waste minimization and ZW objectives should be accomplished (Rodríguez-Guerreiro, Torrijos, & Soto, 2024). Reuse, treatment, and recovery of resources may be accomplished on campus by the university itself or off campus by taking part in local or municipal initiatives. Several wealthy nations have started to implement sustainable waste management programs. The following accomplishments and goals, for instance, matched those of other universities, mostly from the United States (Anacio, 2017):

- The percentage of waste that was diverted or recycled was 53% at Harvard University, 33% at Boston University, 44% at Princeton University, 40% at Appalachian State University, 67% at Rutgers University, and 25% at Arizona State University.
- The percentage of waste that was reduced was 34% at Princeton University and 70% at the Chinese University of Hong Kong.

Effective waste management and sustainability at universities also improve the well-being of students. Stress and mental health problems are common among students who live in substandard conditions and are surrounded by unattractive campuses. In contrast, a clean environment fosters serenity, which makes it easier for students to concentrate on their academics (Shahabuddin, Alam, Krishna, Bhaskar, & Perkins, 2020; Smyth, Fredeen, & Booth, 2010). Table 3.5 lists five revolutionary approaches to efficient waste management on campuses.

Table 3.5 Five revolutionary approaches to efficient waste management on campuses

Approach	*Description*
Waste Fill Level Monitoring	Monitoring bin fill levels prevents bin overflow and unnecessary collections, thereby enhancing the efficiency of trash management on campuses. By using sensors or real-time data to monitor bin capacity, campus personnel can promptly determine when bins need to be emptied. Overflowing trash cans not only make the campus unattractive, but they also provide serious health hazards to students living there by drawing vermin and other disease-carrying animals. The frequency of needless collections may be decreased, which lowers the collection vehicles' carbon footprint and helps control garbage disposal expenses. Campuses may ensure a cleaner, safer, and more enjoyable environment for everyone by having dumpsters adequately inspected and maintained.
Facility Waste Management	In contemporary university trash management, using an app and scanning bin QR codes during collections and disposals have evolved into unique instruments for increasing sustainability and efficiency. Each bin has a QR code that waste management teams may scan to get precise information on the location of the bin, fill levels, and collection dates. Additionally, real-time monitoring of waste metrics, such the weight of the garbage collected and the levels of contamination in recycling bins, is made possible by the use of an app. By keeping track of the weight, campuses may keep an eye on their trash production, gaining insights into trends in waste formation and chances to cut waste in general. These technologies' integration encourages greater data-driven decision-making, which results in better collection routes, lower operating expenses, and a more environmentally friendly campus waste management system.

(*Continued*)

Table 3.5 (Continued)

Approach	*Description*
Bin Access Management	The main benefit of bin access control is that it guarantees that only those with permission may use the bins, which makes remote administration easy and flexible for any kind of bin. This approach is crucial for campus waste management since it makes sure that residents who are in charge of sorting put the right things in the right bins, which helps to minimize contamination. It also prevents illegal dumping, which happens often when trash cans are positioned in public places. Furthermore, bin access management linked to the sensor devices aids in overflow control and maintains a cleaner, more orderly campus.
Rightsizing	Campuses must rightsize their waste infrastructure to maximize efficiency and minimize costs. By using real-time data to monitor the rate at which bins are filling, universities can make informed decisions about the number, capacity, and placement of waste containers, allowing facilities and premises to better align waste management resources with actual usage patterns and preventing dumpsters from overflowing or sitting underutilized, which not only results in unsightly campus conditions but also raises costs due to overage fees and the need for more frequent pickups. Additionally, strategically placing bins in high-traffic areas or redistributing capacity can optimize pick-up schedules, lowering emissions and fuel consumption from waste collection trucks.
Route Planning	Campus garbage collection may benefit from route planning, which increases efficiency by using waste monitoring data. Campuses can better coordinate collection routes to meet the trash infrastructure's real demands. As an example, consider the Sensoneo Driver Navigation App, which enables operators to send drivers planned routes directly. The drivers then get comprehensive itineraries with voice-guided navigation that provides step-by-step instructions. For garbage collection trucks, these routes are designed to be the safest, most accessible, and quickest possible. Through the reduction of needless visits, this improved route design helps to decrease the environmental effect of garbage collection by lowering truck air pollution, noise pollution, and traffic congestion. The software promotes a more orderly and clean campus environment by guaranteeing prompt and effective garbage collections, which enhances the university's overall sustainability and operational effectiveness.

Source: own study based on Sensoneo (2024).

Many buildings, amenities, and infrastructure systems make up a university's usually wide and varied campus (Ismaeil & Sobaih, 2022). Over such intricate networks, controlling water use may be difficult. Several water supply sources, antiquated plumbing systems, and dispersed decision-making procedures are all possible on campuses. A lot of preparation and cooperation across several departments may be needed to coordinate campus-wide water-saving initiatives. Intelligent leak detection technologies and solutions may significantly cut down on water waste from the type of hidden leaks that

Table 3.6 Smart building water management technologies

Technology	*Description*
Water management and monitoring	The water use of a building or campus is precisely recorded and tracked by these systems. Along with identifying inefficiencies, the gathered data often contains insight that suggests places and methods for improving water conservation.
Smart taps and sensors	Water sensors are placed throughout buildings and used in combination with water management systems to measure water accurately and pinpoint places where leaks may be happening or where water is being used. Measurement devices that gather data are also smart taps. Nevertheless, these taps also aid in maximizing use according to the location of building or campus residents.
Humidity and climate sensors	These sensors use weather and humidity readings from the inside and outside to determine the ideal amount of water needed for HVAC and ventilation.
Water treatment and purification	Reusable water that has been cleaned and disinfected by sophisticated filters may assist minimize environmental harm that might jeopardize sustainability initiatives.
Digital twins	Digital twins, which are perhaps the newest technology in the field of water management, model different suggested modifications to campus or building water management procedures to determine the advantages or disadvantages of implementing such changes. As a result, building owners and operators may more precisely adopt sustainability and water-saving measures without having to conduct expensive and time-consuming trial-and-error testing.
Smart irrigation systems	Smart irrigation systems include temperature and humidity sensors to determine when and how much water is needed. This maximizes the watering of green spaces and gardens. Even though it's more often linked to agriculture, smart buildings are already using it since it enhances environmental sustainability, optimizes water usage, and thus lowers energy use.

Source: own study based on Froehlich (2025).

are typical in aged infrastructure. The amount of water a university consumes is greatly influenced by the facilities and courses it provides. Science laboratories, for instance, need a lot of water to run experiments, chill equipment, and keep things hygienic. Water usage is often greater in universities with large sports facilities, swimming pools, and fitness centres. Water must be continuously circulated in pools, and showers in locker rooms increase total demand. Sports fields may also need a lot of irrigation, and the upkeep of the surrounding landscaping may be significant. Similarly, there might be substantial water needs while maintaining decorative plants, lawns, gardens, and trees. These places may significantly reduce their water usage by building stormwater and greywater recycling systems, as well as water-efficient irrigation systems. Water use is often excessive in resident halls and student residences. Kitchen facilities, laundry, and showers are examples of everyday activities that add to the total usage. The key to lowering water use in these regions is making sure that the fixtures are water-efficient and encouraging sustainable behaviours among the inhabitants. Shower periods may be considerably decreased, for instance, by using computerized showers that are programmed to shut off after a certain amount of time. Significant water conservation may also be achieved using waterless urinals, low flow taps, and water-efficient toilets; a single waterless urinal can save up to 100,000 L of water annually. Water use in dining establishments and spaces used for food preparation is high due to cleaning, cooking, and dishwashing. Water usage in these areas may be reduced by the use of water-conserving dishwashers, staff training on water-saving techniques, and efficient kitchen equipment (Li, Chien, Chang, Chou, & Tai, 2022). There are several building verticals where smart water management may be used. These technologies are unique to the efficiency of smart buildings (Table 3.6).

References

Adeniran, A. E., Nubi, A. T., & Adelopo, A. O. (2017). Solid waste generation and characterization in the University of Lagos for a sustainable waste management. *Waste management (New York, N.Y.)*, *67*, 3–10, https://www.sciencedirect.com/science/article/pii/s0956053x17303045

Aghamolaei, R., & Fallahpour, M. (2023). Strategies towards reducing carbon emission in university campuses: A comprehensive review of both global and local scales. *Journal of Building Engineering*, *76*, 107183, https://www.sciencedirect.com/science/article/pii/s2352710223013633

Al-Dmour, H. (2023). Green-smart university campuses: the mediating role of student engagement in enhancing corporate image. *SAGE Open*, *13*(4).

Anacio, D. B. (2017). Designing sustainable consumption and production systems in higher education institutions: the case of solid waste management. In W. Leal Filho, M. Mifsud, C. Shiel, R. Pretorius (Eds.), *Handbook of theory and practice of sustainable development in higher education*. World sustainability series. Cham: Springer. https://doi.org/10.1007/978-3-319-47895-1_1

Apelfröjd, S., Eriksson, S., & Bernhoff, H. (2016). A review of research on large scale modern vertical axis wind turbines at Uppsala University. *Energies*, *9*(7), 570, https://www.mdpi.com/1996-1073/9/7/570

Ardito, L., Ferraris, A., Messeni Petruzzelli, A., Bresciani, S., & Del Giudice, M. (2019). The role of universities in the knowledge management of smart city projects. *Technological Forecasting and Social Change*, *142*, 312–321, https://www.sciencedirect.com/science/article/pii/s0040162517317754

Basic Information | Green Building |US EPA (2016). Retrieved January 03, 2025, https://archive.epa.gov/greenbuilding/web/html/about.html

Carlson, S., Wagner, R., Korpi, J., Caswell, D., Karmaker, L., & Snow, M. (2022). *How much and how much better? A case study of zero-net energy housing in Northern Minnesota.*

Chen, H., Jiang, W., Yang, Y., Yang, Y., & Man, X. (2017). State of the art on food waste research: A bibliometrics study from 1997 to 2014. *Journal of Cleaner Production*, *140*, 840–846.

Cross, Stan, Eagan, D. J., & Tolmé, P., with Keniry, J., & John, K. (2011). *Going underground on campus: Tapping the earth for clean, efficient heating and cooling, a guide to geothermal energy and underground buildings on campus*, Foreword by Robert J. Koeste, National Wildlife Federation.

Dahlawi, S., & El Sharkawy, M. F. (2021). Assessment of solid waste management practice in the university campus. *International Journal of Sustainability in Higher Education, 22*(3), 561–575, https://www.emerald.com/insight/content/doi/10.1108/ijshe-05-2020-0183/full/pdf

Dai, Y., Liu, A., & Lim, C. P. (2023). Reconceptualizing ChatGPT and generative AI as a student-driven innovation in higher education. *Procedia CIRP, 119*, 84–90, https://www.sciencedirect.com/science/article/pii/s2212827123004407

Dong, Z. Y., Zhang, Y., Yip, C., Swift, S., & Beswick, K. (2020). Smart campus: definition, framework, technologies, and services. *IET Smart Cities, 2*(1), 43–54.

Dowling, C. B., Neumann, K., & Florea, L. (2016). *Geothermal energy: An important resource*: Geological Society of America.

Estep, J. (n.d.) *Technoeconomic analysis of geothermal district heating in the Boston, MA area*. Massachusetts Institute of Technology, https://dspace.mit.edu/handle/1721.1/157136

Froehlich, A. (NaN). Water conservation technology is key to tomorrow's smart buildings. *Buildings*. Retrieved January 10, 2025, https://www.buildings.com/resiliency-sustainability/energy-water-efficiency/article/33037072/water-conservation-technology-is-key-to-tomorrows-smart-buildings

Gholami, H., Bachok, M. F., Saman, M. Z. M., Streimikiene, D., Sharif, S., & Zakuan, N. (2020). An ISM approach for the barrier analysis in implementing green campus operations: Towards higher education sustainability. *Sustainability, 12*(1), 363, https://www.mdpi.com/2071-1050/12/1/363

Global approaches to sustainability through learning and education (2019). IGI Global Scientific Publishing, IGI Global.

Gursoy Haksevenler, B. H., Kavak, F. F., & Akpinar, A. (2022). Separate waste collection in higher education institutions with its technical and social aspects: A case study for a university campus. *Journal of Cleaner Production, 367*, 133022, https://www.sciencedirect.com/science/article/pii/s0959652622026142

Huertas, J. I., Mahlknecht, J., Lozoya-Santos, J. d. J., Uribe, S., López-Guajardo, E. A., & Ramirez-Mendoza, R. A. (2021). Campus city project: Challenge living lab for smart cities. *Applied Sciences, 11*(23), 11085, https://www.mdpi.com/2076-3417/11/23/11085

Ismaeil, E. M. H., & Sobaih, A. E. E. (2022). Assessing Xeriscaping as a retrofit sustainable water consumption approach for a Desert University Campus. *Water, 14*(11), 1681.

Jakimiuk, A., Matsui, Y., Podlasek, A., Koda, E., Goli, V. S. N. S., Vobĕrková, S., et al. (2023). Closing the loop: A case study on pathways for promoting sustainable waste management on university campuses. *The Science of the Total Environment, 892*, 164349, https://www.sciencedirect.com/science/article/pii/s0048969723029704

Kanelli, A. A., Vardaka, M. L., Malesios, C., Katima, Z. J., & Kalantzi, O.-I. (2024). Can campus green spaces be restorative? A case study from Tanzania. *Sustainability, 16*(3), 1094, https://www.mdpi.com/2071-1050/16/3/1094

Karvonen, A., Martin, C., & Evans, J. (2018). University campuses as testbeds of smart urban innovation. In *Creating smart cities* (pp. 104–118). Routledge.

Kelly, O., Illingworth, S., Butera, F., Dawson, V., White, P., Blaise, M., et al. (2022). Education in a warming world: Trends, opportunities and pitfalls for institutes of higher education. *Frontiers in Sustainability, 3*, 1–8.

Li, P., Chien, H., Chang, P., Chou, S., & Tai, C.-H. (2022). Water management strategies on campus: An integrated approach. *Journal of Sustainability Perspectives, 2*(1), 73–79.

Li, R., Zhao, R., Xie, Z., Xiao, L., & Chuai, X. (2025). Water–energy–carbon nexus among different functional areas of campus: Pattern, flow, and efficiency. *Sustainable Cities and Society*, *118*, 106039.

Li, Y., Chen, H., & Yu, P. (2024). Green campus transformation in smart city development: A study on low-carbon and energy-saving design for the renovation of school buildings. *Smart Cities*, *7*(5), 2940–2965.

Majeed, A., & Ali, M. (2018). How Internet-of-Things (IoT) making the university campuses smart? QA higher education (QAHE) perspective. In *2018 IEEE 8th annual computing and communication workshop and conference (CCWC)* (pp. 646–648). IEEE.

Makhloufi, L., Vasa, L., Rosak-Szyrocka, J., & Djermani, F. (2023). Understanding the impact of big data analytics and knowledge management on green innovation practices and organizational performance: The moderating effect of government support. *Sustainability*, *15*(11), 8456, https://www.mdpi.com/2071-1050/15/11/8456

Moghayedi, A., Michell, K., Hübner, D., Le Jeune, K., & Massyn, M. (2024). Examine the impact of green methods and technologies on the environmental sustainability of supportive education buildings, perspectives of circular economy and net-zero carbon operation. *Facilities*, *42*(3/4), 201–222, https://www.emerald.com/insight/content/doi/10.1108/f-12-2022-0161/full/pdf

Momani, K. M. K., Nour, A.-N. I., & Jamaludin, N. (2020). Sustainable universities and green campuses. In *Global approaches to sustainability through learning and education* (pp. 17–27). IGI Global Scientific Publishing, IGI Global.

Oluwunmi, A. O., Oladayo, O. P., Role, B. A., & Afolabi, T. O. (2019). Benefits and barriers to the implementation of green building standards in universities: what are students' views? *IOP Conference Series: Materials Science and Engineering*, *640*(1), 12031.

Omotayo, T., Moghayedi, A., Awuzie, B., & Ajayi, S. (2021). Infrastructure elements for smart campuses: A bibliometric analysis. *Sustainability*, *13*(14), 7960.

Omotayo, T., Olanipekun, A., Obi, L., & Boateng, P. (2020). A systems thinking approach for incremental reduction of non-physical waste. *Built Environment Project and Asset Management*, *10*(4), 509–528.

Pupiales-Chuquin, S.-A., Tenesaca-Luna, G.-A., & Mora-Arciniegas, M.-B. (2022). Proposal of a methodology for the implementation of a smart campus. *Proceedings of sixth international congress on information and communication technology*, *236*, 589–602, https://link.springer.com/chapter/10.1007/978-981-16-2380-6_52

Ragazzi, M., & Ghidini, F. (2017). Environmental sustainability of universities: critical analysis of a green ranking. *Energy Procedia*, *119*, 111–120, https://www.sciencedirect.com/science/article/pii/s1876610217325961

Rodríguez-Guerreiro, M.-J., Torrijos, V., & Soto, M. (2024). A review of waste management in higher education institutions: The road to zero waste and sustainability. *Environments*, *11*(12), 293.

Sensoneo (2024). *5 strategies for effective waste management on university campus*. Retrieved January 10, 2025, https://sensoneo.com/waste-library/waste-management-university-campus/

Shahabuddin, M., Alam, M. T., Krishna, B. B., Bhaskar, T., & Perkins, G. (2020). A review on the production of renewable aviation fuels from the gasification of biomass and residual wastes. *Bioresource Technology*, *312*, 123596.

Shaheen, F., Lodhi, M. S., Rosak-Szyrocka, J., Zaman, K., Awan, U., Asif, M., et al. (2022). Cleaner technology and natural resource management: An environmental sustainability perspective from China. *Clean Technologies*, *4*(3), 584–606, https://www.mdpi.com/2571-8797/4/3/36

Shankar, Y. S., & Khandelwal, R. (2017). Sustainable waste management strategy for a campus: a case study of JUET, Guna. *Management of Environmental Quality: An International Journal*, *28*(5), 610–623, https://www.emerald.com/insight/content/doi/10.1108/meq-01-2016-0008/full/pdf

Smart education and learning market. *Global report [2027]* (2024). Retrieved December 03, 2024, https://www.fortunebusinessinsights.com/industry-reports/smart-education-and-learning-market-101942

Smart university: A university in the technological age (2021). *TEM Journal, 10*(1), 13–17, https://www.ceeol.com/search/article-detail?id=935160

Smyth, D. P., Fredeen, A. L., & Booth, A. L. (2010). Reducing solid waste in higher education: The first step towards 'greening' a university campus. *Resources, Conservation and Recycling, 54*(11), 1007–1016.

Teah, H. S., Yang, Q., Onuki, M., & Teah, H. Y. (2019). Incorporating external effects into project sustainability assessments: The case of a green campus initiative based on a solar PV system. *Sustainability, 11*(20), 5786, https://www.mdpi.com/2071-1050/11/20/5786

Technology management and its social impact on education. (2023). IGI Global Scientific Publishing; IGI Global.

Thakur, M. (2007). The impact of ranking systems on higher education and its stakeholders. *Journal of Institutional Research, 13*(1), 83–96, https://eric.ed.gov/?id=ej1055590

Thomashow, M. (2014). *The nine elements of a sustainable campus.* MIT Press.

Toutouh, J., & Alba, E. (2022). A low cost IoT cyber-physical system for vehicle and pedestrian tracking in a smart campus. *Sensors, 22*(17), 6585, https://www.mdpi.com/1424-8220/22/17/6585

Weijermars, R., Burnett, D., Claridge, D., Noynaert, S., Pate, M., Westphal, D., et al. (2018). Redeveloping depleted hydrocarbon wells in an enhanced geothermal system (EGS) for a university campus: Progress report of a real-asset-based feasibility study. *Energy Strategy Reviews, 21*, 191–203, https://www.sciencedirect.com/science/article/pii/s2211467x18300488

Wind resource data, tools, and maps. *Geospatial Data Science | NREL* (2025). Retrieved January 10, 2025, https://www2.nrel.gov/gis/wind

Wlda (2022). *Top 10 solar universities 2022 | YSG solar | YSG | World-leading solar farm & battery energy storage development firm.* Retrieved January 09, 2025, https://www.ysgsolar.com/blog/top-10-solar-universities-2022-ysg-solar/

4 Smart Learning Environments

4.1 Digital Classrooms and Learning Management Systems

As of now, there's no one strict definition as to what is smart learning. Overall, smart education is a new educational concept that provides and expands the whole process of learning outcome and certification in digital age. The English acronym for **'SMART'** further divides the word into the following categories according to its individual letters: **S – self-directed, M – motivated, A – adaptable, R – resource-enriched, and T – technological** (Ardashkin, 2018). Smart learning is also described as 'an emerging area alongside other related emerging areas like smart technology, smart teaching, smart education, smart-e-learning, smart classrooms, smart universities, and smart society' (2025).

The primary goal of smart learning is to provide students with the most recent knowledge in education and technology, as well as the essential skills they need to study more effectively, professionally, flexibly, happily, and resourcefully in order to fulfil the demands of contemporary society. Furthermore, these learning strategies provide a completely contextualized learning process to help students with specific learning, living, and working situations. In addition to allowing the student to develop their abilities, this special contextualized learning process enhances the quality of their instruction (Ţălu, 2020).

Improving the traditional learning approach is critical. In this regard, the legal and educational authorities have undertaken a number of initiatives. Both the technology and educational worlds have been rapidly impacted by the rapid development of smart education technologies and automated learning management systems in recent years (Fardinpour, Pedram, & Burkle, 2014; Srivastava, Tiwari, Kumar, & Garg, 2024). The use of cutting-edge gear and software in the educational process, for example, as well as the wide range of virtual platforms available in contemporary classrooms enables the creation of dynamic, interactive educational experiences that support personalized, flexible learning and student collaboration. With its intelligent and virtual instruments to facilitate learning, the creative learning management system may be regarded as a strong mechanical learning model (Hoel & Mason, 2018) (Figure 4.1).

With the use of digital technology in education, the current educational environment has improved or changed for the better (Haleem, Javaid, Qadri,

DOI: 10.4324/9781003621829-5

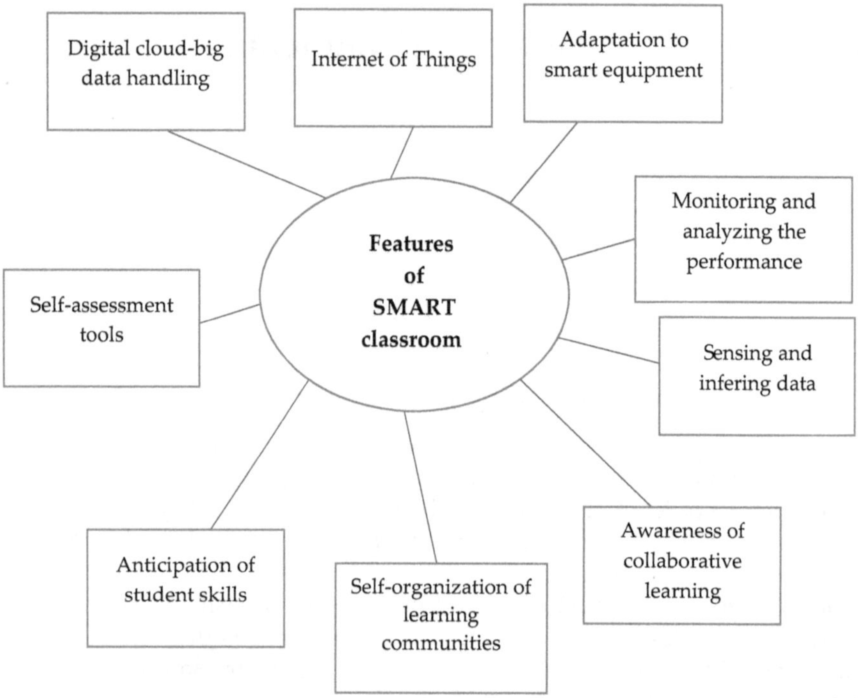

Figure 4.1 Features of smart classroom.

Source: own study based on Iqbal, Parra-Saldivar, Zavala-Yoe, & Ramirez-Mendoza (2020).

& Suman, 2022; Rosak-Szyrocka, 2024; Rosak-Szyrocka, Żywiołek, Nayyar, & Naved, 2023; Santos et al., 2022; Singh, 2021). Because it extends the classroom beyond the four walls into the community, the digital classroom is beneficial to all students. Social media, multimedia, and mobile phones are examples of electronic devices or platforms that are used to educate students in digital classrooms (Haleem et al., 2022). They help students and educators communicate, collaborate, and share knowledge, creating a sense of community (Haleem et al., 2022; Singh, 2021).

4.2 Features of Smart Classroom

Students make use of devices such as laptops, tablets, Chromebooks, and other internet-connected devices that are essential for accessing, creating, and sharing digital educational materials, fostering connectivity and interactivity. Digital tools enable personalized education, tailored to the way an individual student learns, his or her interests, and pace. Students engage in actual projects of real-life situations. Through an interesting and dynamic platform, the majority of the curriculum is presented to students online, eliminating the need for them to take notes on what the instructor has taught. We are able to divide the online classroom into two halves. First, every student in the synchronous digital classroom has a computer, and those who attend virtually may use the Internet to participate in class. The instructor uses the computer to educate using cutting-edge technology and oversee the learning process. In addition, asynchronous digital classes allow all students to engage online at any time and from any location. With the use of online learning tools, this student-centred approach to learning enables a network of learners to share knowledge across time and location barriers (Haleem et al., 2022; Rasulova, 2024). Table 4.1 shows elements of a digital classroom.

Digital technologies in education have given rise to various learning management systems (LMSs). These LSMs have promoted virtual classrooms where a

Table 4.1 Elements of digital classroom

Element	*Description*
E-books	E-books allow students to access knowledge quickly and easily. Students do not have to carry paper books. E-books are compatible with various devices such as smartphones, laptops, and tablets, so students can acquire knowledge regardless of their location. Another advantage of e-books is that they are cheaper than paper books, and many of them have open access, that is, free access. In e-books, students can search for keywords or fragments that interest them without having to browse the entire book, as is the case with paper books. E-books can also change the font or the background colour, something particularly useful for students with vision issues or that are sleepy at night. Many e-book platforms allow students to easily share notes, highlights, and insights with other students, engaging them in collaborative learning. Digital libraries store an entire library of books at a single location and occupy no physical space.
Online learning	Online learning enables flexible and remote access to educational content through virtual platforms, making education accessible to students regardless of location (Grek, 2008; Nguyen, Zierler, & Nguyen, 2011; Roschelle, Sharples, & Chan, 2005; Villagrasa, Fonseca, Redondo, & Duran, 2014).
Personalized learning	On its own, artificial intelligence is able to determine a student's strengths and shortcomings. To guarantee that every student realizes their full potential, they use a customized learning approach. Students with physical or learning limitations may benefit from assistive technology, which enables them to engage fully in class and efficiently assimilate topics. Innovative technologies for the visually impaired include speech recognition, screen-reading tools, Braille displays, and text-to-speech solutions; for the hearing impaired, closed-captioning apps, sound amplifiers, and video conferencing technologies enable lipreading and sign language.
Augmented reality (AR)/ virtual reality (VR)	Students may experience topics of study directly and interactively by using VR to go to whole new locations or historical eras. AR assists students in understanding difficult subjects by providing them with an immersive visual representation of abstract concepts (Haleem et al., 2022).

(*Continued*)

Table 4.1 (Continued)

Element	*Description*
Open Educational Resources (OER) 5Rs	OER means freely accessible and openly licensed educational materials that enhance learning opportunities while reducing costs for students and institutions. The UNESCO (UnescoPhysicalDocument 2002) definition of OER is the use of information and communication technology by a learning group to embrace non-commercial study materials. Free and accessible OER must be made possible by information and communication technologies. The content must be technically, financially, and legally available, with minimal restrictions on copyright, license, and cost in order to be considered open (Nikoi & Armellini, 2012; Open educational resources: Opportunities and challenges 2006). It must be possible for the user to modify and personalize OER content to suit their requirements (Frango Silveira, 2016). User retention, reuse, revision, remix, and redistribution must all be allowed for OER content. The 5Rs are the aggregate name for these applications.
Social media platforms	Messaging allows students and teachers to have real-time conversations with one another, in video calls or group chats, enabling them to clear queries, share insights, and discuss any topic any time. These platforms help learning go beyond the limits of a lecture or a classroom. For example, private groups on platforms like Facebook keep classes together by providing updates, announcements, and follow-ups on academic work. Through regular communication, a collaborative environment is established where students feel heard and supported.

Source: own study based on Haleem et al. (2022).

teacher can interact with students in real-time, share his resources, deliver his lecture, assess students' learning, collect feedback, and reply to their queries (Barakina, Popova, Gorokhova, & Voskovskaya, 2021; Mason, 2006). In a conventional LMS, an instructor or moderator prepares and delivers material, keeps track of student involvement, and evaluates students' performance online. Students may use interactive elements offered by the LMS. Therefore, an LMS's primary features include video conferencing, threaded conversations, and discussion forums. Monitoring student growth and performance is the aim of an LMS. The LMS, being a web-based learning platform, makes it possible to access learning materials and administration at 'any time, any place, any pace' (Learning management system (LMS) among university students: Does it work, 2013). Moodle, WebCT, Canvas, Blackboard, and Desire2Learn are a few LMSs that are used in educational institutions (Frango Silveira, 2016; Iqbal et al., 2020; Nikoi & Armellini, 2012). Key Features of LMS are the following (Mpungose & Khoza, 2022; Sulaiman, 2023):

- Content management is the creation and administration of educational resources, modules, and courses.
- Constant supervision of teaching effects and monitoring of students' progress in learning.
- Using tools such as quizzes to assess students' knowledge.

Table 4.2 Benefits of LMS in education

Benefit	*Description*
Enhanced accessibility and flexibility	The course materials are available to students from anywhere in the globe at any time. Higher completion rates and improved information retention result from this flexibility, which lets students interact with the training material at their own speed.
Scalability	The majority of LMSs can easily handle an increasing number of users and courses without posing serious logistical or financial difficulties. Universities' training programs may increase in tandem with their growth thanks to this scalability.
Centralized learning environment	Every educational resource, student file, and evaluation outcome is kept in one single area. This ensures consistency of education delivery across the institution and ensures smooth running in administration tasks.
Personalization and adaptability	Modern LMSs offer personalized learning paths tailored to the needs and preferences of individual users. Adaptive learning features, such as adjusting the pace and difficulty based on the performance of the learner, may enhance the efficiency of a learning platform and improved engagement.
Engagement through gamification	Gamification through elements such as badges, medals, or gaining points increases the joy of learning and at the same time increases the student's motivation and involvement in learning.
Advanced reporting and analytics	LMS offers detailed analytics on performance data, completion rates, and learning habits. These analytics can empower the institution to make data-driven decisions in order to continually improve its educational offerings over time.

Source: own study.

- LMS allows to obtain a report on current trends in education.
- Using communication tools such as forums.

Table 4.2 shows benefits of LMS in education.

4.3 Virtual and Augmented Reality in Education

Virtual reality (VR) and augmented reality (AR) are two of the most cutting-edge technological developments available today, and they have enormous potential to enhance the educational system (Kamińska et al., 2019). Despite their similar names, AR and VR are two independent technologies with different applications (Tan, Xu, Li, & Chen, 2022; Young, Stehle, Walsh, & Tiri, 2020). One of the definitions of VR is that it is ***I3* = *Interaction* + *Immersion* + *Imagination*** (Burdea & Coiffet, 2003). VR is now mostly produced by using head-mounted display (HMD) devices to generate visual effects. An HMD is a head-mounted or helmet-mounted device with integrated lenses and a display that lets the user experience the virtual world via a broad field of vision, tracking of hand and head motions, and controller-controlled object interaction (Zhang, 2017). Learning and teaching processes may be supported and facilitated by VR.

The majority of students recalled what they saw in VR, according to several polls and publications, and they came to the conclusion that VR is a more memorable setting than lab-based demos (Teeroumanee Nadan, Alexandrov, Jamieson, & Watson, 2011; Slavova, 2018). The technology known as AR overlays computer-generated content on the real world around us to improve it (Hantono, Nugroho, & Santosa, 2018). AR integrates digital information, such as text, images, videos, and 3D objects, into the real world (Tan et al., 2022). Using a variety of devices, including computers, mobile cameras, and webcams, together with a variety of software, AR is a novel technology that blends the real and virtual worlds. This allows for the presentation of pictures that seem to be objects, including people, animals, objects, and spacecraft, as 3D objects hovering above the real surface. In order to produce AR, the actual and virtual worlds were combined (Pellas, Mystakidis, & Kazanidis, 2021). Moreover, AR may be integrated into VR headsets that support pass-through mode by displaying data from built-in camera sensors. As a result, spectators will feel as if they are taking part in a 'real' event (Rosak-Szyrocka, Żywiołek, & Shahbaz, 2023).

There are many chances to take use of technology-enhanced learning thanks to the growing usage of AR and VR in education in recent years (Tan et al., 2022). Students can better engage with complex material beyond lectures and textbooks thanks to AR and VR, which expose them to immersive digital experiences that are impossible to replicate through traditional teaching methods (Gupta & Rosak-Szyrocka, 2024; Rosak-Szyrocka, Żywiołek, & Shahbaz, 2023). Additionally, lecturers can tailor content to each student's unique learning style (Childso et al., 2023). These technologies have the ability to not only make the experience more immersive but also to allow instructors to build simulations and take students on virtual field excursions without the need for actual travel (Seidametova, Abduramanov & Seydametov, n.d.). Furthermore, cutting-edge technologies like AR and VR may help close the gap between conventional classroom learning and real-world experience, offering measurable advantages for students' professional growth. VR transports users to a whole new virtual world, offering an interactive experience via glasses or headgear. By superimposing digital items on top of the actual world, adding more information to it, or improving its functioning, AR enhances the real world. A more engaging learning environment and immersive learning experiences are only two advantages that AR and VR technologies provide to online, mobile, and hybrid learning. With the use of AR and VR, virtual worlds and simulations may be created that let students explore and engage with real-world settings without ever leaving the classroom (Young, Stehle, Walsh, & Tiri, 2020). In order to keep students interested, AR and VR may also be utilized to provide interactive and captivating material, such as 3D pictures and films (Gargrish, Mantri, & Kaur, 2020; Sun, Ye, Yu, & Chiu, 2023). The usage of AR and VR technology may make the material being taught more relevant. These technologies may improve students' comprehension of the material being taught by enabling them to explore various settings and engage with it in a more realistic manner. These technologies may also be utilized to more engagingly present new ideas and concepts, which can boost students' excitement and engagement. Students with special needs may benefit from a more accessible educational experience thanks to AR and VR technology. These technologies may help people with physical or cognitive limitations learn more easily by letting them experience virtual worlds on their own terms and at their own speed (Al-Ansi, Jaboob, Garad, & Al-Ansi, 2023). Table 4.3 shows case studies of pioneering universities in using VR and AR.

When analysing the issue of VR and AR, it is worth paying attention to opportunities and challenges (Table 4.4).

Table 4.3 Case studies of pioneering universities in using VR and AR

University	*Country*	*Description*
Arizona State University (ASU)	United States	ASU's usage of VR in its online biological science degree program shows how the technology may be used to overcome geographical limitations and enable distant students to take part in lab activities electronically.
University of Maryland	United States	By combining the digital and physical worlds, this university uses AR to improve anatomy education for students.
McMaster University	Canada	By developing simulations, interactive visualizations, and virtual labs, McMaster's use of VR in engineering education demonstrates how immersive technology may provide innovative learning experiences.
Stanford University	United States	As an early adopter of VR for immersive education, Stanford School of Business has been providing a diploma program that is wholly taught via VR.
Queen Mary University of London	United Kingdom	To improve medical education and training, medical students at this institution have attended lectures in 'the metaverse', which uses VR.
Grinnell College, Iowa	United States	To promote a better comprehension of a variety of courses, students at Grinnell College collaborate on VR development projects that include historical recreations, such as Viking longship, and immersive renderings of mathematical ideas.
Harvard University	United States	Students utilize VR to study ancient Egyptian characters along a tomb as part of the Giza Project, a joint endeavour that uses immersive technology to improve their knowledge of archaeology and anthropology.
Zhejiang University	China	As part of their partnership with Harvard University for the Giza Project, Zhejiang University students learn with their Harvard counterparts in VR-enabled classrooms, demonstrating the global cooperation that VR can foster.
University of Southern California (USC)	United States	USC offers a distinctive approach to experiencing and comprehending history outside of standard textbooks by using VR to immerse students in historical situations.

Source: own study.

Table 4.4 Opportunities and challenges of VR and AR in university

Aspect	*Opportunities*	*Challenges*
Engagement and interaction	AR and VR make communication, motivation, and interaction between peers, advisers, and learners possible while providing immersive interactive learning.	Younger or less tech-savvy students might feel overwhelmed or struggle to engage effectively with AR and VR.
Learning effectiveness	This technology enables virtual field trips, interactive lessons, and even replicating real-world experiences; enhancing students' understanding and retention of concepts	Not every student or institution has or can afford the hardware and software needed to take full advantage of this access.
Cost efficiency	Lowers the expenses on material for learning and travelling, thus enabling learning to be cheaper and more achievable.	High upfront costs for hardware, software, and even ongoing updates make it difficult for schools with tight budgets to adopt AR and VR.
Personalization	Provides individualized education through personalized lessons and assessments for improved student outcomes and engagement.	Scalability is a major problem: schools need to keep updating content and technologies offered to stay in line with developments in AR and VR, and this can be resource-intensive.
Collaboration	Impacts collaboration among students and between students and teachers in virtual spaces, promoting teamwork and peer-to-peer learning.	A shortage of well-qualified teachers trained in the effective application of AR and VR can limit their use for collaborative and immersive learning.
Gamification	Enhances motivation and engagement by integrating gamified features in lessons making learning interesting.	Gamified content needs to be regularly updated to keep consistent with new educational standards superseding older methods.
Assessment tools	Use interactive and personalized quizzes that gather users understanding and offer feedback when needed.	Some schools do not have the technological infrastructure to support VR- and AR-based assessment.
Accessibility	Breaks down geographical and physical barriers to provide learning from anywhere.	In remote or underprivileged areas, the limitations in terms of hardware and connectivity make AR and VR in education less accessible.
Innovation	With AR and VR, teachers can create lessons that are dynamic, interactive, and highly visual, keeping students engaged while improving learning outcomes.	It takes a tremendous amount of resources to keep up with the development/implementation of advanced technologies.

Source: own study.

4.4 Personalized and Adaptive Learning Systems

Higher education is changing as a result of this paradigm shift from conventional instructor-centred to student-centred methods. When learning environments are tailored to the profiles of the learners, they become more personalized, improving both the learning experience and the learners' performance. 'Instruction in which the pace of learning and the instructional approach are optimized for the needs of each learner' is the definition of personalized learning given in the 2017 US National Education Technology Plan (Xie, Chu, Hwang, & Wang, 2019). Personalized learning involves adapting teaching strategies to each student. The development of personalized learning supports the notion that, in recognition of the variety of learners, students learn best when teaching is tailored to their specific requirements (Dockterman, 2018; Ryoo & Winkelmann, 2021). Customizing student learning is made possible by the concept of personalized learning, which allows for a change in the way that learning is designed from a 'one size fits all' paradigm to an adaptable and student-centred approach (Hidayat & Afuan, 2021). As a result, learning is maximized, helping students quickly and affordably accomplish course goals (Raj & Renumol, 2022), providing education that meets the requirements of students of all ages (Burak & Gultekin, 2024).

Maintaining consistency in the delivery of content is essential for any learning environment with a specific goal in order to increase student motivation to learn (Ilić, Mikić, Kopanja, & Vesin, 2023). This ensures that students receive individualized content specific to their needs rather than generic, one-size-fits-all material (Arsovic & Stefanovic, 2020). These kinds of settings may encourage self-learning, which attracts students and increases their involvement in the educational process (El-Sabagh, 2021).

If a learning environment can monitor user activities, interpret them using domain-specific models, deduce user needs and preferences from the interpreted activities and appropriately represent them in related models, and, finally, act upon the knowledge that is currently available about its users and the subject matter at hand, it is said to be adaptive. Personalized learning is made possible by adaptive learning technology, which evaluates learners' existing knowledge and skills, provides feedback and content, and continuously tracks progress using learning algorithms that offer real-time updates and the resources needed to enhance student learning (EDUCAUSE Library, 2025).

Based on data collected by an automated system, the new personalized learning method makes judgements automatically, adapts to learners' real-time learning environments, and ensures that learning activities and material match each learner's unique requirements and features (Peng, Ma, & Spector, 2019). Developments in adaptive e-learning systems have enabled and aided personalized learning, which is education that is customized to a student's unique needs and learning preferences (Beldagli & Adiguzel, 2010).

Adaptive learning is an educational approach that utilizes technology, particularly machine learning algorithms, to provide personalized learning experiences tailored to individual students' needs, preferences, and progress. Machine learning algorithms are used by adaptive learning systems to collect, process, and evaluate enormous volumes of learner data. By adapting to the specific requirements of each student, adaptive learning promotes effective and efficient learning, maximizes engagement, and enhances educational outcomes (El-Sabagh, 2021; Gligorea et al., 2023). In order to ensure that students get the most pertinent and successful instructional materials and activities, adaptive learning systems may provide timely and focused interventions via ongoing monitoring of learner data, including assessment outcomes, interaction patterns, and progress tracking (Gligorea et al., 2023). Adaptive learning gives students the possibility to become aware of their own learning speed (Dry, Due, Powell, Chur-Hansen, & Burns, 2018), which allows them to progress at their own rate and close the learning gap with their classmates (Feng, Cui, & Wang, 2018). They also foster the development of autonomous learning abilities (Augmenting formative writing assessment with learning analytics: A design abstraction approach, 2018). With the help of this data-driven method, the system can adapt the learning pathway and provide individualized materials, exercises, and topics that are tailored to the abilities and objectives of each student. Self-paced learning, tailored help, and a more productive and interesting learning environment are all encouraged by adaptive learning. These tools are able to find trends in student data, pinpoint areas of strength and weakness, and provide tailored suggestions and interventions. Instructors and designers may also improve and optimize the e-learning environment by using the adaptive learning technique to get useful feedback and data on the efficacy of instructional learning tactics and resources (Gligorea et al., 2023). Table 4.5 shows the characteristic of personalized and adaptive learning systems.

There are two types of adaptive learning technologies: adaptive learning programs and adaptive learning platforms. While adaptive learning platforms are standalone systems that combine all courseware features into a single, reliable unit, adaptive learning programs provide a component that can be integrated into an existing course on an LMS to create a customized and adaptable experience. The adaptive systems may provide material created by the vendor or they may just act as a structure for institutions to enter their own content. Adaptive learning technology has a lot of intriguing potential to improve student learning and achievement (*Adaptive learning systems: Surviving the storm*, 2016). A publisher, institution, or vendor's content may be supported by adaptive learning systems. The teacher and the institution must weigh the advantages and disadvantages of each alternative, taking into account institutional resources, availability, content quality, and technological usability. The methods of adaptation also differ. There are four types of adaptive learning systems available at the moment, and they all include learning goals, teaching materials, and evaluations into learning modules. The four frameworks are

Table 4.5 Characteristic of personalized and adaptive learning systems

Aspect	*Personalized learning systems*	*Adaptive learning systems*
Aim	Involves adapting teaching and learning materials and approaches according to the preferences, interests, and objectives of the student.	Using algorithms and data from student performance to automatically adjust content and difficulty level in real time. From user performance and behaviour, the system automatically generates learning paths to support user learning process. Ensures proposed material and learning is at the right level – not too easy nor too difficult, optimizing the learning process.
Control	Have a good amount of freedom in choosing their learning streams and skills.	The process is done based on algorithms and data analytics
Customization	It is student or teacher specific and targets pre-defined preferences and aspirations.	These changes can be made dynamically based on the student's current behaviours and development.
Technology dependence	Can operate with little real-time data processing but might have to be set up manually by teachers.	Relying mainly on advanced algorithms, AI, and machine learning for rapid adaptation.
Feedback	Gives general feedback according to settings or what the teacher told.	Provides instant and highly contextualized feedback based on a student's activities and performance.
Example	Platforms that allow students to decide what topics, goals, or learning styles to pursue, for example, Coursera.	Systems, such as DreamBox or ALEKS, which tailor difficulty based on students' real-time responses to tasks and questions.

Source: Own study.

machine learning–based adaptive systems, rule-based adaptive systems, advanced algorithm–based adaptive systems, and decision tree adaptive systems, in order of complexity. Some of the new adaptive technologies being utilized in higher education are **Cengage Mindtap, Knewton, CogBooks, SmartSparrow, and Realizeit**. To provide feedback, scaffolding, and continuous assessment that deliver the tailored learning routes, powerful computers process assessment, interaction, and learning behaviour data for each student. This allows for the large-scale personalization of learning. Because formative assessments continually monitor and evaluate students' performance in the course, adaptive learning systems may provide teachers with information early in the course, even before the first test. An at-risk student may be identified before it's too late by using early intervention techniques and sending information to the teacher, which will lower attrition and boost student achievement. 'Adaptive systems reduce faculty load in teaching and remediation to teaching and facilitating, as well as course content progression based on students' skill and outcomes mastery measurement' (*Adaptive learning systems: Surviving the storm*, 2016). They also handle the fundamentally diverse degrees of previous knowledge.

The adaptive learning system presents the adaptive information in a sequence that is tailored to each student after tracking performance and behaviours during engagement. Multiple alternatives for the information delivery mode, such as text, video, or interactive activities, are provided by a strong adaptive learning system. Certain adaptive systems are able to assess a student's performance, determine whether they learn best via interactive exercises, viewing videos, or reading, and then provide the material in the most effective way possible in relation to their increased performance (Alrawashdeh, Fyffe, Azevedo, & Castillo, 2024; Taylor, Yeung, & Bashet, 2021). These systems may also accomplish personalization by evaluating learner performance or engagement without identifying individual qualities or preferences (Xie, Chu, Hwang, & Wang, 2019). These systems may take the form of artificial intelligence (AI)-driven solutions or operate based on if-then conditional rules (limited algorithms). For instance, Squirrel AI (created in China) utilizes over 10 machine learning algorithms to adjust language, math, and science material, along with activity sequencing, to correspond with students' requirements, interests, and skills to achieve mastery (*2022 AAAI squirrel AI award for artificial intelligence for the benefit of humanity*, 2021). Conversely, OneTab (United Kingdom) applies restricted algorithms to give individualized reading material (Adam, 2024). These systems allow for continuous and flexible adaptation, and learner progress is estimated by updating the ability estimates either automatically using AI algorithms, as in Squirrel AI, or frequently at predetermined intervals using conditional rules, as in the case of the Kenyan application M-Shule, which adapts math and English learning content using a series of If-then rules (Adam, 2024). A combination of both can be used, such as in the expert system Mangahigh (United Kingdom), which identifies what a student already knows, how well they know it, and what they need to learn next by tracking their performance on teacher-assigned tasks, peer performance, and national curriculum expectations in math (Adam, 2024). Many American universities have implemented adaptive learning systems, and China is adopting them at a faster pace. In the last three years, the number of unique users for one product alone has approached 2 million. Given China's quick adoption, it is helpful to look at how effective adaptive learning is in that nation's educational system (Wang et al., 2023). Table 4.6 shows case studies of universities using personalized and adaptive learning systems.

Table 4.6 Case studies of universities using personalized and adaptive learning systems

University	*System type*	*Platform/ technology*	*Description/results*
Arizona State University (ASU) United States	Adaptive	ALEKS (Assessment and Learning in Knowledge Spaces)	ASU uses adaptive learning platforms to create individualized math and science courses, adjusting difficulty according to a student's pace in the material. Results: Increased pass rates and decreased dropouts in introductory courses
University of Michigan (United States)	Adaptive	ECoach	Students have the ability to shape their own learning pathways with flexible course options available wherever possible. Results: Increase in student engagement and improved outcomes in STEM courses
Georgia State University (United States)	Adaptive	iPASS (Integrated Planning and Advising for Student Success)	Monitors students' academic progress and provides real-time interventions to address potential concerns. Results: Decreased time-to-graduation and increased retention rates.
Stanford University (United States)	Personalized	Socratic and Canvas LMS	Personalized learning for graduate-level courses, providing customized resources based on student goals Results: Enhanced learning productivity for students with a wide range of academic and career goals.
Nanyang Technological University (Singapore)	Adaptive	AI-driven LMS and LingoLab	Give systems that adapt based on student behaviour in language and technical skills training. Results: Improved achievement in technical skills and linguistic ability.
University of Central Florida (United States)	Personalized and adaptive	Realizeit	Hybrid and online courses that combine personalized learning and adaptive technology. Results: Improved course completion rates and narrower gaps in learner outcomes.

(*Continued*)

Table 4.6 (Continued)

University	*System type*	*Platform/ technology*	*Description/results*
University of New South Wales (Australia)	Personalized and adaptive	Moodle LMS with Analytics Integration	Utilizes adaptive analytics that monitor student progress and deliver personalized content recommendations. Results: Increased interactions in online learning settings and better completion rates.
University of Edinburgh (United Kingdom)	Personalized and adaptive	Moodle with Learning Analytics	Integrates adaptive analytics and personalized pathways on online and blended learning programs. Results: Improved student satisfaction and progress in modular courses.
University of Queensland (Australia)	Personalized and adaptive	edX and Smart Sparrow	Combines adaptive accelerators for skill-building with custom-mapped coursework for individuals' goals Results: Higher retention in MOOCs and enhanced results in continued education programs.
Deakin University (Australia)	Personalized and adaptive	FutureLearn with Adaptive Features	Employs adaptive platforms that deliver personalized learning paths for online and hybrid programs. Results: Higher retention rates and a better learning experience for remote and working students.
Delft University of Technology (Netherlands)	Personalized and adaptive	edX and Adaptive MOOC Platforms	Creates customized MOOCs with adaptive elements and personalized learning programs for technical disciplines. Results: Higher engagement of students across the globe in specialized engineering courses.
Lund University (Sweden)	Personalized and adaptive	Moodle and AI-Based Learning Tools	Provides customized digital courses with dynamic testing and individualized assistance for all students. Results: For lifelong learning programs better engagement and learning outcomes
University of Hong Kong (HKU)	Personalized and adaptive	AI-Powered LMS and Analytics Dashboard	Uses AI-based systems to customize content and assessments according to students' cultural and educational context. Results: More inclusive and improved satisfaction among international students

Source: own study.

References

2022 AAAI squirrel AI award for artificial intelligence for the benefit of humanity (2021). https://www.computer.org/csdl/api/v1/periodical/mags/co/2021/04/download-issue/pdf

Adam, I. (2024). Development and efficacy of adaptive personalised learning environments: A systematic review and meta-analysis. *Pakistan Journal of Life and Social Sciences (PJLSS)*, *22*(2), 7223–7244.

Adaptive learning systems: Surviving the storm (2016).

Al-Ansi, A. M., Jaboob, M., Garad, A., & Al-Ansi, A. (2023). Analyzing augmented reality (AR) and virtual reality (VR) recent development in education. *Social Sciences & Humanities Open*, *8*(1), 100532. https://www.sciencedirect.com/science/article/pii/s2590291123001377

Alrawashdeh, G. S., Fyffe, S., Azevedo, R. F., & Castillo, N. M. (2024). Exploring the impact of personalized and adaptive learning technologies on reading literacy: A global meta-analysis. *Educational Research Review*, *42*, 100587. https://www.sciencedirect.com/science/article/pii/s1747938x23000805

Ardashkin, I. B., Chmykhalo, A., Makienko, M. A., & Khaldeeva, M. A. (2018). Smart-technologies in higher engineering education: Modern application trends. *European Proceedings of Social and Behavioural Sciences*, *50*, 57–64. https://www.europeanproceedings.com/article/10.15405/epsbs.2018.12.8

Arsovic, B., & Stefanovic, N. (2020). E-learning based on the adaptive learning model: Case study in Serbia. *Sādhanā*, *45*(1), 1–13. https://link.springer.com/article/10.1007/s12046-020-01499-8&casa_token=iajox8ogxysaaaaa:8luvikuqt5cosdnyp8inucz3kr7wcghgt5mtbc4bbhd7scbi3iuv-llnbelksjafx0f_lrzebuo9pq

Augmenting formative writing assessment with learning analytics: A design abstraction approach (2018). https://repository.isls.org/handle/1/856

Barakina, E. Y., Popova, A. V., Gorokhova, S. S., & Voskovskaya, A. S. (2021). Digital technologies and artificial intelligence technologies in education. *European Journal of Contemporary Education*, *10*(2), 285–296. https://eric.ed.gov/?id=ej1311498

Beldagli, B., & Adiguzel, T. (2010). Illustrating an ideal adaptive e-learning: A conceptual framework. *Procedia - Social and Behavioral Sciences*, *2* (2), 5755–5761. https://www.sciencedirect.com/science/article/pii/s1877042810009791

Burak, D., & Gultekin, M. 2024, Implementation and evaluation of an adaptive learning environment designed according to learner characteristics: A study on primary school social studies teaching. *Technology, Knowledge and Learning*, *29*(1), 163–196. https://link.springer.com/article/10.1007/s10758-022-09623-9&casa_token=ileklxbx9taaaaaa:kxdgv9b3ofdfstu3gjlpc98c5qzhfme8rmtezbcrb0msl5v8tq9ceoubcbv5qaup27x6rhkpffrnvq

Burdea, G. C., & Coiffet, P. (2003). *Virtual reality technology*. John Wiley & Sons.

Childso, E., Mohammad, F., Stevens, L., Burbelo, H., Awoke, A., Rewkowski, N., & Manocha, D. (2023). An overview of enhancing distance learning through augmented and virtual reality technologies. https://arxiv.org/ftp/arxiv/papers/2101/Z2101.11000.pdf

Dockterman, D. (2018). Insights from 200+ years of personalized learning *npj Science of Learning*, *3*(1), 15. https://www.nature.com/articles/s41539-018-0033-x

Dry, M. J., Due, C., Powell, C., Chur-Hansen, A., & Burns, N. R. (2018). Assessing the utility of an online adaptive learning tool in a large undergraduate psychology course. *Psychology Teaching Review*, *24*, (2), 24–37. https://eric.ed.gov/?id=ej1196465

EDUCAUSE Library (2025). *7 Things you should know about adaptive learning*. https://library.educause.edu/resources/2017/1/7-things-you-should-know-about-adaptive-learning [12 January 2025].

El-Sabagh, H. A. (2021) Adaptive e-learning environment based on learning styles and its impact on development students' engagement. *International Journal of Educational*

Technology in Higher Education, *18*(1), 1–24. https://link.springer.com/article/10.1186/s41239-021-00289-4

Fardinpour, A., Pedram, M. M., & Burkle, M. (2014). Intelligent learning management systems. *International Journal of Distance Education Technologies*, *12*(4), 19–31. https://www.igi-global.com/article/intelligent-learning-management-systems/121737

Feng, M., Cui, W., & Wang, S. (2018). *Adaptive learning goes to China* (pp. 89–93). Cham: Springer. https://link.springer.com/chapter/10.1007/978-3-319-93846-2_17

Frango Silveira, I. (2016). OER and MOOC: The need for openness. *Issues in Informing Science and Information Technology*, *13*, 209–223.

Gargrish, S., Mantri, A., & Kaur, D. P. (2020) Augmented reality-based learning environment to enhance teaching-learning experience in geometry education. *Procedia Computer Science*, *172*, 1039–1046. https://www.sciencedirect.com/science/article/pii/s1877050920314824

Gligorea, I., Cioca, M., Oancea, R., Gorski, A.-T., Gorski, H., & Tudorache, P. (2023). Adaptive learning using artificial intelligence in e-learning: A literature review. *Education Sciences*, *13*, (12), 1216.

Grek, S. (2008). From symbols to numbers: The shifting technologies of education governance in Europe. *European Educational Research Journal*, *7*(2), 208–218.

Gupta, S. K., & Rosak-Szyrocka, J. (2024). *Innovation in the University 4.0 system based on smart technologies*. Boca Raton: Chapman and Hall/CRC.

Haleem, A., Javaid, M., Qadri, M. A., & Suman, R. (2022). Understanding the role of digital technologies in education: A review. *Sustainable Operations and Computers*, *3*, 275–285. https://www.sciencedirect.com/science/article/pii/s2666412722000137

Hantono, B. S., Nugroho, L. E., & Santosa, P. I. (2018). Meta-review of augmented reality in education. In *2018 10th international conference on information technology and electrical engineering (ICITEE)* (pp. 312–315). IEEE.

Hidayat, N., & Afuan, L. (2021). Naïve Bayes for detecting student's learning style using felder-silverman index. *JUITA: Jurnal Informatika*, *9*(2), 181.

Hoel, T., & Mason, J. (2018). Standards for smart education – towards a development framework. *Smart Learning Environments*, *5*(1), 1–25. https://link.springer.com/article/10.1186/s40561-018-0052-3

Ilić, M., Mikić, V., Kopanja, L., & Vesin, B. (2023). Intelligent techniques in e-learning: A literature review. *Artificial Intelligence Review*, *56*(12), 14907–14953. https://link.springer.com/article/10.1007/s10462-023-10508-1

Iqbal, H. M. N., Parra-Saldivar, R., Zavala-Yoe, R., & Ramirez-Mendoza, R. A. (2020). Smart educational tools and learning management systems: Supportive framework. *International Journal on Interactive Design and Manufacturing (IJIDeM)*, *14*(4), 1179–1193. https://link.springer.com/article/10.1007/s12008-020-00695-4&casa_token=96-7jkymbtgaaaaa:jmlj5w3rruqp5mucxkptaoxa6e_owfz-vroa2xkrqpbdk4ts_tfu0mtuhteyhrjwmusvqiuw-ffupq

Kamińska, D., Sapiński, T., Wiak, S., Tikk, T., Haamer, R., Avots, E., Helmi, A., Ozcinar, C., & Anbarjafari, G. (2019). Virtual reality and its applications in education: Survey. *Information*, *10*(10), 318. https://www.mdpi.com/2078-2489/10/10/318

Learning management system (LMS) among university students: Does it work (2013). https://www.researchgate.net/profile/nor-adzharuddin/publication/269838611_learning_management_system_lms_among_university_students_does_it_work/links/5ad20396458515c60f50a322/learning-management-system-lms-among-university-students-does-it-work.pdf

Mason, R. (2006). Learning technologies for adult continuing education. *Studies in Continuing Education*, *28*(2), 121–133.

Mpungose, C. B., & Khoza, S. B. (2022). Postgraduate students' experiences on the use of moodle and canvas learning management system. *Technology, Knowledge and Learning*, *27*(1), 1–16. https://link.springer.com/article/10.1007/s10758-020-09475-1

Nadan, T., Alexandrov, V., Jamieson, R., & Watson, K. (2011). Is virtual reality a memorable experience in an educational context?. *International Journal of Emerging Technologies in Learning (iJET)*, *6*(1), 53–57. https://www.learntechlib.org/p/44953/

Nguyen, D. N., Zierler, B., & Nguyen, H. Q. (2011). A survey of nursing faculty needs for training in use of new technologies for education and practice. *Journal of Nursing Education*, *50*(4), 181–189.

Nikoi, S., & Armellini, A. (2012). The OER mix in higher education: Purpose, process, product, and policy. *Distance Education*, 33(2), 165–184.

Pellas, N., Mystakidis, S., & Kazanidis, I. (2021). Immersive virtual reality in K-12 and higher education: A systematic review of the last decade scientific literature. *Virtual Reality*, *25*(3), 835–861. https://link.springer.com/article/10.1007/s10055-020-00489-9

Peng, H., Ma, S., & Spector, J. M. (2019). Personalized adaptive learning: An emerging pedagogical approach enabled by a smart learning environment. *Smart Learning Environments*, *6*(1), 1–14. https://link.springer.com/article/10.1186/s40561-019-0089-y

Raj, N. S., & Renumol, V. G. (2022). A systematic literature review on adaptive content recommenders in personalized learning environments from 2015 to 2020. *Journal of Computers in Education*, 9(1), 113–148. https://link.springer.com/article/10.1007/s40692-021-00199-4

Rasulova, T. P. (2024). *Impacts of digital classrooms in education*. https://libeldoc.bsuir.by/handle/123456789/54981

Rosak-Szyrocka, J. (2024). The era of digitalization in education where do Universities 4.0 go? *Management Systems in Production Engineering*, *32*(1), 54–66. https://sciendo.com/article/10.2478/mspe-2024-0006

Rosak-Szyrocka, J., Żywiołek, J., Nayyar, A., & Naved, M. (2023). *The role of sustainability and artificial intelligence in education improvement*, CRC Press.

Rosak-Szyrocka, J., Żywiołek, J., & Shahbaz, M. (2023). *Quality management, value creation, and the digital economy*. London: Routledge.

Roschelle, J., Sharples, M., & Chan, T. W. (2005). Introduction to the special issue on wireless and mobile technologies in education. *Journal of Computer Assisted Learning*, *21*(3), 159–161.

Ryoo, J., & Winkelmann, K. (2021). *Innovative learning environments in STEM higher education*. Cham: Springer International Publishing. https://library.oapen.org/handle/20.500.12657/47325

Santos, G., Magalhaes, M. D., Carvalho, S., Pinto, R. J., Felix, M. J., & Rosak-Szyrocka, J. (2022). Digital transformation and global inequality. *International Journal for Quality Research*, *16*(4), 1297–1314.

Seidametova, Z. S., Abduramanov, Z. S. & Seydametov, G. S. (n.d.). Using augmented reality for architecture artifacts visualizations. *1613-0073*. https://elibrary.kdpu.edu.ua/handle/123456789/4626

Singh, M. N. (2021). Inroad of digital technology in education: Age of digital classroom. *Higher Education for the Future*, *8*(1), 20–30.

Slavova, Y., & Mu, M. (2018). A comparative study of the learning outcomes and experience of VR in education. In *2018 IEEE conference on virtual reality and 3D user interfaces (VR)*. IEEE.

Srivastava, A., Tiwari, S., Kumar, D., & Garg, N. (2024). Finding of DDoS attack in IoT-based networks using ensemble technique. In *2024 International Conference on Intelligent Systems for Cybersecurity (ISCS)* (pp. 1–4), IEEE.

Sulaiman, T. T. (2023). A systematic review on factors influencing learning management system usage in Arab gulf countries. *Education and Information Technologies*, *29* (2), 1–19. https://link.springer.com/article/10.1007/s10639-023-11936-w

Sun, J. C.-Y., Ye, S.-L., Yu, S.-J., & Chiu, T. K. F. (2023). Effects of wearable hybrid AR/VR learning material on high school students' situational interest, engagement, and learning performance: The case of a physics laboratory learning environment.

Journal of Science Education and Technology, *32*(1), 1–12. https://link.springer.com/article/10.1007/s10956-022-10001-4&casa_token=k-ejj07vrleaaaaa:j3zqsc_mw2n-a1prqj-1mnmnocd5nght8elaldawuyrnvgh2umbpuqcilibc5e1chzsu11o6ikgawa

Ţălu, Ş. (2020). New perspectives in the implementation of smart-technologies in higher education. In *2nd international scientific and practical conference "Modern Management Trends and the Digital Economy: From Regional Development to Global Economic Growth" (MTDE 2020)* (pp. 253–257). https://www.atlantis-press.com/proceedings/mtde-20/125939718

Tan, Y., Xu, W., Li, S., & Chen, K. (2022). Augmented and Virtual Reality (AR/VR) for education and training in the AEC industry: A systematic review of research and applications. *Buildings*, *12*(10), 1529. https://www.mdpi.com/2075-5309/12/10/1529

Taylor, D. L., Yeung, M., & Bashet, A. Z. (2021). Personalized and adaptive learning. In *SpringerBriefs in statistics* (pp. 17–34), Cham: Springer International Publishing.

Villagrasa, S., Fonseca, D., Redondo, E., & Duran, J. (2014). Teaching case of gamification and visual technologies for education. *Journal of Cases on Information Technology*, *16*(4), 38–57. https://www.igi-global.com/article/teaching-case-of-gamification-and-visual-technologies-for-education/120703

Wang, S., Christensen, C., Cui, W., Tong, R., Yarnall, L., Shear, L., & Feng, M. (2023). When adaptive learning is effective learning: Comparison of an adaptive learning system to teacher-led instruction. *Interactive Learning Environments*, *31*(2), 793–803.

Xie, H., Chu, H.-C., Hwang, G.-J., & Wang, C.-C. (2019). Trends and development in technology-enhanced adaptive/personalized learning: A systematic review of journal publications from 2007 to 2017. *Computers & Education*, *140*, 103599. https://www.sciencedirect.com/science/article/pii/s0360131519301526

Young, G., Stehle, S., Walsh, B., & Tiri, E. (2020). Exploring virtual reality in the higher education classroom: Using VR to build knowledge and understanding. *JUCS - Journal of Universal Computer Science*, *26*(8), 904–928.

Zhang, H. (2017). Head-mounted display-based intuitive virtual reality training system for the mining industry. *International Journal of Mining Science and Technology*, *27*(4), 717–722. https://www.sciencedirect.com/science/article/pii/s2095268617303439

5 Integrating Smart Mobility and Transportation on Campus

5.1 Autonomous Transportation and Campus Shuttles

Due to their shape, population, demand, and patterns, college campuses have particular mobility needs. It is often advised to reduce the number of private vehicles on campuses since they may disrupt research and instruction, degrade the campus's aesthetics owing to parking, pollute the environment, and have detrimental health impacts (Kim et al., 2017). By 2050, the usage of self-driving autonomous vehicles (AVs) may cut greenhouse gas emissions from transportation by up to 34% (Ercan et al., 2022). (Iclodean, Cordos, & Varga, 2020). Vehicle automation may be divided into six stages, Level 0 through Level 5, in accordance with the Society of Automotive Engineers (SAE) standard. The degree of human involvement needed to complete a job is the main basis for the categorization (Figure 5.1). The field of autonomous driving has undergone a revolution since 2000. Once this technology is implemented widely, the short- and medium-term benefits of autonomous driving will clearly affect public transit. Levels 3 and 4 of the autonomy were attained by the shuttle buses now operating on public highways (Iclodean et al., 2020). The rapid advancement of technology in this area is shown by the emergence of autonomous taxis and bus services (An, 2024). In San Francisco, businesses like Alphabet's Waymo and GM's Cruise are setting the standard for Level 4 autonomous services (Waymo, 2025). Similar state-of-the-art services are also being provided by Baidu in China, demonstrating the enormous potential of autonomous driving (2025). Furthermore, pilot projects in the Sangam area of South Korea demonstrate that this tendency is widespread and that efforts are being made to bring AVs to the world.

Since DARPA Grant Challenge contests, when many institutions created automobiles with autonomous driving capabilities around confined circuits, the development of self-driving cars has accelerated (Block, Keller, & Negoita, 2024). The development of AVs began at the academic level and has since spread to the industrial level. As a result, commercialized vehicles now come equipped with various forms of autonomy that help drivers with driving, parking, and safety.

DOI: 10.4324/9781003621829-6

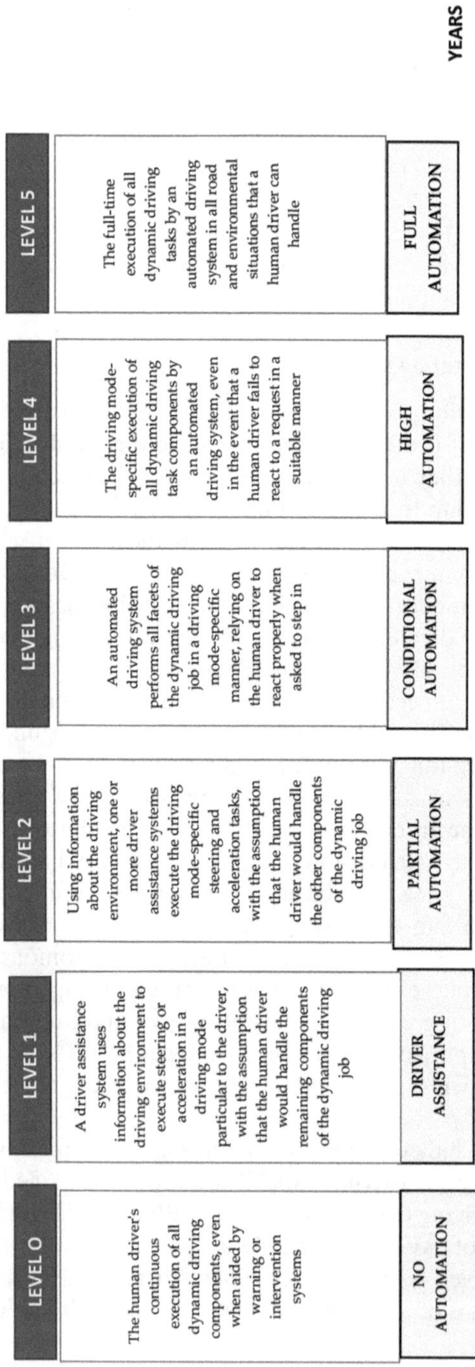

Figure 5.1 Levels of vehicle autonomy.

Sources: own study based on Autonomous Campus Mobility Platform, (2018), Iclodean et al. (2020), and Ondruš, Kolla, Vertaľ, & Šarić (2020).

In order to provide accessibility and parking, universities provide on-campus shuttle services for employees and students. Recently, several colleges, including the University of Michigan, have implemented AV shuttles on campus (Nicefaro, 2019). These AV shuttles might provide mobility services at athletic events and other special occasions, such as graduation ceremonies, in addition to transferring students to class facilities without the hassle of parking (*How do college students perceive future shared mobility with autonomous Vehicles? A survey of the University of Alabama students*, 2022). College football has played a significant role in the University of Alabama campus culture. On game days at the University of Alabama, SAVs may also be crucial in reducing traffic jams and relieving strain on the campus parking lot (Fu, Alabama, Liu, & Jones, 2022; Zhang, Guhathakurta, Fang, & Zhang, 2015). Two instances of shared automation pilot projects are the fully hybrid electric Lexus vehicles used in downtown Arlington and on the University of Texas Arlington campus (2025) and a shuttle that was tested in March 2019 using Level 4 self-driving vehicles in the West Village (WV) neighbourhood of the University of California, Davis (UCD) campus (*Exploring the role of attitude in the acceptance of self-driving shuttles*, 2020).

The structure, population density, and spatiotemporal movement patterns of college campuses provide special transportation needs. Due to disruptions in research and teaching, environmental pollution, parking-related visual deterioration, and negative health consequences on employees and students, private vehicle commuting has to be curtailed. The first and final mile issue affects campus shuttle systems just as it does for other public bus networks. Particularly on big, hilly, or rural campuses, bike and pedestrian-friendly rules are not a universally applicable answer. As an alternative, a driverless taxi promotes car use and offers point-to-point mobility while lowering private vehicle commute costs. Public transportation systems, such as transit buses and shuttle services, reduce traffic congestion, pollutants, fuel consumption, the danger of traffic accidents, and the general loss of productivity that goes along with them. Complimentary shuttle or bus services are a vital form of transportation on campus and in the surrounding areas for the majority of big and mid-size open-campus colleges. Automated shuttle service has been identified as a possible approach to ease the following issues: traffic congestion on campus, traffic congestion between classes, short-distance excursions inside or around campus, and trouble locating parking places. Peer high-tech businesses like EasyMile, NAVYA, and Meridian have built or are developing automated shuttle/bus systems to travel short distances and predetermined routes in facilities with multiple uses (Table 5.1).

The state-of-the-art autonomous shuttle/bus systems that are now available have been successfully demonstrated in a variety of situations and are equipped with navigation, route planning and control, obstacle recognition, and/or accident-avoidance capabilities. The majority of autonomous shuttle testing and investment has historically taken place in Europe; two recent experiments were carried out on college campuses: one at Wageningen University in the

Table 5.1 Automated shuttle vendor technologies

Shuttle	*Description*
EasyMile (EZ10)	EasyMile is a leading global provider of autonomous technologies and smart mobility solutions. It creates software that automates different transportation platforms, offers a robust in-house fleet management system for AVs, and offers smart mobility solutions for passenger transportation in both public and private, urban and rural environments. The most popular locations have been theme parks and cultural venues, city centres, retirement homes, and college campuses. EasyMile is the only wheelchair-accessible autonomous shuttle. Without a steering wheel or pedal, it is entirely electric, doesn't need any extra equipment, and can be connected to smart infrastructure (like traffic signals). The shuttle uses a mix of cameras, differential GPS, and LIDAR for navigation. The bus's sensors gather and process recorded data to provide a 360-degree view of the surroundings, including any obstacles in the way, other cars, pedestrians, and infrastructure. The autonomous car system can make decisions about how to proceed safely on the road by analysing data in real-time (stop, go, slow down, etc.). A variety of sensors are used by driverless technology to perceive its surroundings. The embedded mechanism in EasyMile's proprietary software package automates transportation platforms. It gathers and analyses recorded data from a wide variety of sensors to provide a comprehensive image of the environment around AVs. Three essential elements primarily guarantee tasks: Localization: What location is the shuttle?, Perception: What's going on around the shuttle? Path Planning: What's the next move for the shuttle?. Z10 always knows their location and its precision thanks to a fusion algorithm that uses all the data from the many sensors. The shuttle will safely slow down or stop if there is any deviation. Furthermore, the vehicle maintains continuous connection with the EasyMile oversight centre and its surroundings. By combining data from the sensor set, EasyMile's robust localization algorithm determines the vehicle's location in real-time with centimetre-level precision. The vehicle's broadest field of view and low ground clearance are guaranteed by a comprehensive array of high-performance sensors. As a result, it can recognize and steer clear of such roadblocks. It will either adjust its speed and trajectory, safely apply the brake, or come to a halt, depending on the circumstances.

NAVYA AUTONOM	With its electric, driverless, and autonomous solutions, NAVYA helps cities and private locations all over the globe meet their mobility demands. Based in Lyon, France, NAVYA began developing autonomous shuttles in 2015 with its initial offering, ARMA. The AUTONOM shuttle was created by NAVYA to provide both autonomous transport performance and a pleasant first- and last-mile journey. With its efficient guiding and detection systems that integrate many technologies, such as LIDAR, GPS, odometer, and stereovision camera, AUTONOM operates without a steering wheel or pedals. Obstacle detection and 3D mapping are done with LIDAR, vehicle position is done with GPS, speed is determined with an odometer, and environmental analysis and obstacle detection are done with a camera. Salford-Manchester (UK), Gothenburg (Sweden), Michigan (United States), and Adelaide (Australia) are among the colleges that have already made the most of their transportation offerings using NAVYA's cutting-edge technology. The majority of these foreign campuses are private. Therefore, no special permissions are needed to deploy autonomous shuttles. NAVYA can hold a maximum of 15 people.
OLLI	Using many micro-factories, Local Motors is an American automaker that specializes in producing open-source car designs in small quantities. Local Motors was established in 2007 and made its debut in the driverless shuttle market with OLLI, an electric autonomous shuttle. Constructed in collaboration with IBM, it has IBM Watson technology to provide passengers a customized experience. Additionally, it is the first driverless shuttle presently produced using additive manufacturing methods like 3D printing. The shuttle's many sensors allow it to view in all directions, and its built-in software recognizes items like automobiles, pedestrians, and cyclists. It also employs advanced algorithms to determine the speed and trajectory of other vehicles and make proactive navigational judgments. Olli can hold a maximum of 20 people. Speed limit: 25 km/h.
COAST AUTONOMOUS SHUTTLE P1	Coast Autonomous (COAST) is a software and technology firm that combines a range of cars with AV technology to provide Transportation-as-a-Service (TaaS) solutions to cities, theme parks, campuses, airports, and other low-speed locations. The specialists on the Pasadena, California-based COAST team were the leaders of the only European team competing in the DARPA Grand Challenges in 2005 and 2007. Its first self-driving, low-speed shuttle, created in 2010, has successfully conducted 60 demonstrations throughout seven nations. Vehicle makers and transport operators that prioritize mobility in low-speed conditions (up to 25 mph) may benefit from COAST's full automation. The P1 is a bi-directional vehicle that can carry people throughout the day thanks to its stereo cameras, GPS, LIDAR sensors, and V2X technologies. In order to train cars to learn the whole network of potential pathways inside the environment, the COAST system first 3D maps the entire area where the vehicle operates. The COAST software turns the vehicle into a driverless shuttle to provide route-based or on-demand service when it has learned the surroundings.

Sources: own study based on COAST (2025), Lin et al. (2018), Lin, Menon, Chen, & Kourtellis (2020), and Medrano-Berumen, Malayjerdi, Akbas, & Sell (2020).

Netherlands in 2016 and one at the Swiss Federal Institute of Technology in Lausanne, Switzerland, in 2015. Autonomous shuttle testing or demonstration has gained interest in the United States in the last two years. Around 1.3 million people use the institution of South Florida's (USF) Bull Runner bus service each year across the Tampa campus and surrounding areas. USF is a prominent US institution with around 50,000 students enrolled (Campus automated shuttle service deployment initiative, 2018). Table 5.2 shows case studies of universities with autonomous mobility programs/initiatives.

Table 5.3 shows opportunities and challenges for autonomous transportation on campuses.

Table 5.2 Case studies of universities with autonomous mobility programs/initiatives

University	*Program/ initiative*	*Description*	*Achievements*
University of Michigan (United States)	MCity Shuttle Program	Autonomous shuttle pilot program over fixed route on a closed course simulating urban and campus settings.	Successfully tested autonomous navigating shuttles, data used to improve autonomous systems.
University of Gothenburg (Sweden)	Autonomous Campus Mobility	Piloted interlinked public transport autonomous shuttles with local organizations.	Used electric AVs to promote green transportation and strengthen campus connectivity.
Arizona State University (United States)	Orbit Shuttles	Launched self-driving shuttles to carry students on brief campus rides, as part of sustainability objectives.	Decreased vehicle emissions on campus; and improved transportation options for students with disabilities.
University of Adelaide (Australia)	Smart Mobility Project	Enrolled in a smart city program that includes AVs testing on campus.	Enhanced industry collaboration; better assimilation of autonomous systems into urban environments.
Texas A&M University (United States)	Autonomous Freight and Shuttles	The campus launched a program for autonomous shuttles and goods delivery vehicles.	Shuffled and repackaged autonomous transport to cover logistics; increased operational efficiency Lower costs for autonomous shuttles vs traditional logistics solutions.

(*Continued*)

Table 5.2 (Continued)

University	*Program/ initiative*	*Description*	*Achievements*
Singapore University of Technology and Design (Singapore)	Intelligent Campus Transit System	Used advanced AI to test autonomous buses looping between campus hubs.	Promoted high-level AI integration in human transport: established models for urban use in a Singapore context.
École Polytechnique Fédérale de Lausanne (EPFL), Switzerland	SmartShuttle System	A driverless electric shuttle service linking multiple destinations across the site in a shared mobility environment.	Improved integrated autonomy and safety, contributed to sustainability initiatives. Handling shared roads with cars, bicycles, and pedestrians, safety in mixed-traffic environments

Source: own study.

Table 5.3 Opportunities and challenges for autonomous transportation on campuses

Opportunities	*Challenges*
Enhanced Efficiency: Self-driving cars can calculate the fastest routes, streamlining transit and enhancing accessibility.	**Infrastructure Upgrades**: Needs changes in roads, parking, and charging stations to accommodate new technologies.
Improved Safety: Removal of human error, reliance on advanced safety systems.	**Expense**: Cost of initial deployment and upkeep
Reducing Carbon Emissions and Achieving Sustainability Goals: A large number of AVs are electric.	**Legal Compliance**: Issues related to local, state, and federal regulations governing autonomous agents.
Accessibility Solutions: Assist people with disability or mobility problem.	**Public Perception**: Establishing user confidence in the security and dependability of autonomous tech.
Long-Term Cost Savings: Though there may be upfront costs, lower labour costs and increased operational efficiency may make this investment null over time.	**Technical Issues**: Rewards of system malfunctions, cybersecurity threats, and data privacy concerns.
Innovation and Prestige: Showcases a university's investment in cutting-edge technology and sustainability, driving student and staff attraction.	**Integration Obstacles**: Making sure they do compatible with present open transportation frameworks and grounds framework.

Source: own study.

5.2 Sustainable Mobility Options and Public Transit Integration

Mobile sources may emit CO_2, however, on-road sources account for the great bulk of mobile source emissions. These include personal automobiles and campus shuttles that transport administrators, staff, academics, and students to and from the institution as well as during business trips. Along with emissions from on-road cars, business travel by airline and nonroad equipment like lawnmowers, leaf blowers, and small maintenance vehicles used on campus also contribute significantly to emissions (Clifford & Cooper, 1995).

A significant contributor to greenhouse gas (GHG) emissions is the transportation sector; in the United States, for instance, transportation accounted for around 29% of all GHG emissions in 2019 – more than any other sector (US EPA, 2015). Transportation accounts for around 25% of all greenhouse gas emissions in Canada (Saha & Fatmi, 2021). The commute, which is mostly made by automobile, accounts for a significant amount of transportation emissions (Pérez-Neira, Rodríguez-Fernández, & Hidalgo-González, 2020). In this regard, universities may become advocates via climate change policies and programs. In its Climate Action Plan 2030, for instance, the University of British Columbia's Okanagan (UBCO) campus aims to reduce greenhouse gas emissions from the transportation sector by 50–60% by 2030 compared to 2007 levels (Climate Action, 2025). To lower its carbon footprint, UBCO is now creating its mobility master plan and assessing aggressive GHG reduction techniques such flexible work and learning schedules and parking fees (2025). For example, free campus shuttle buses, paid parking permits, bike infrastructure, and other measures may cut down on single-occupancy automobile trips, which will significantly lower emissions (Logan, Nelson, Osbeck, Chapman, & Hastings, 2020). The next major shift in the mobility space seems to be ride-sharing, electrification, and autonomous cars. It may result in fewer traffic, less parking, and safer roadways (Salonen & Haavisto, 2019).

Transportation Demand Management (TDM) strategies must be incorporated into comprehensive transportation system plans for university campuses because fostering a multimodal and effective transportation system on campus will be the primary means of changing graduates' transit attitudes and behaviours for eventual transfer of strategies into their communities and daily lives (Gava & Willaschek, 2023). TDM was introduced as part of a suite of executive solutions to address issues in the transportation industry. TDM promotes management techniques to foster more efficient and ecologically responsible transportation attitudes. The art of altering transportation behaviour is its definition. Numerous advantages of TDM strategies include reduced energy consumption, environmental and natural resource preservation, effective land use, fewer traffic accidents and congestion, a decrease in pollution, more transportation options, and an overall improvement in liveability and social equity (Latif, Chowdhury, & Bagchi, 2024). Additionally, TDM solutions need to strike a balance between the incentive to reduce private vehicle trips and travel choice (Bond & Steiner, 2011). A well-known TDM tactic is the carpool and vanpool program, which gives single-occupancy private vehicle users the chance to eliminate the need for separate car journeys. According to this tactic, two or more persons may utilize a single car if they have the same source, route, and destination (Erdoğan, Cirillo, & Tremblay, 2015). Typically, this approach includes individuals who work and reside in the same or adjacent communities (Balsas, 2003). Fuel and parking expenses must be split by joint users of this technique. For journeys that public transit does not adequately support, this approach is the most popular one. People who commute from outlying locations with fewer public transit options will find the carpool program particularly helpful in small towns and rural areas. On university campuses, which are distinct and homogeneous societies, this strategy may also be highly helpful. Among its numerous concrete and intangible advantages are fuel and parking cost savings, time savings, increased social contact possibilities, stress reduction, congestion and pollution reduction, energy conservation, and support for a healthier future (Erdoğan et al., 2015).

Among the most often used TDM techniques for college campuses are public transit initiatives such as the U-Pass Program. In order to encourage all participants to use public transportation (such as buses, trains, or light rail) and/or active transportation (such as bicycles, walking, etc.) instead of driving their own cars, the U-Pass Program subsidizes the cost of public transportation and makes it more accessible to more potential users (Hickman, Lopez, Cao, Lira, & Biona, 2018b). The U-Pass approach has been successful in reducing the demand for parking spaces on college campuses and boosting the number of people using public transit (Azzali & Sabour, 2018b). The top five reasons for implementing the U-Pass Program, according to American university officials, are: reduced parking demand and traffic; better access to housing and the campus for all members; lower travel and student education costs; greater transportation justice; and higher levels of use of sustainable, active modes of transportation (Hickman et al., 2018b). Additionally, the U-Pass program offers universities a number of benefits, including assisting universities in fulfilling their environmental obligations; reducing the need for parking lots, which frees up more land for educational purposes; and reducing parking spaces and the impact of traffic on the surrounding area (Azzali & Sabour, 2018c). The advantages that U-Pass programs provide to the local community, such as lowering motorized vehicle trips, increasing physical activity, decreasing traffic, and limiting air and noise pollution, are significant for university-community relation (Hickman, Lopez, Cao, Lira, & Biona, 2018a). The quantity and position of shelters, their quality, illumination, seating components, route/schedule information, signs, scheduling, and other affective aspects all have an impact on the quality of services when it comes to encouraging the use of public transit (Bond & Steiner, 2011). The third biggest university in Canada and one of the main sources of traffic in the Vancouver area, the University of British Columbia (UBC), developed an example of a U-Pass Program (Toor & Havlick, 2004). Making campuses more walkable and encouraging the use of bicycles is one tactic colleges use to achieve a more sustainable transportation system (Azzali & Sabour, 2018a).

According to a Turo study, over 50% of young Millennials are choosing not to buy a vehicle for budgetary concerns (2025). With the worldwide market for shared mobility predicted to reach \$1.2 trillion in five years, this generation's increasing dependence on smartphones is opening up new options. By 2028, for instance, the worldwide automobile keyless entry industry is expected to grow to \$3 billion, and 80% of Gen Z currently books travel and rental experiences on their phones, preferring mobile check-ins. Table 5.4 presents mobility options for university campuses mobility.

Table 5.4 Mobility options for university campuses mobility

Mobility option	*Description*
Sharing services	One way of lowering carbon emissions is mass transit on campus. It comes in many different shapes, such as carpooling or car sharing. Students and employees can use a vehicle from a campus fleet when necessary. Carpooling lets two or more individuals share a ride to a common destination, reducing commute costs by 60%.
Biking	One of the most common methods of alternative transportation on campus is biking, requiring little investment but extensive health benefits. Implementing this type of transport needs investments in multiple sectors: • Assessment of infrastructure and coverage of biking lanes; • Installation of racks; • Anti-theft systems on campus cycling stations, electric bike charging points; • Educational campaigns describing the benefits of using a bicycle; • Showers and changing rooms; • Discounts on bike repair and maintenance for students and staff; • Riding free or cheap bike-sharing programs on campus; • Everything under the umbrella of branded renting or sharing of biking equipment; • Bike repair stations. The University of Louisville, for instance, got on board with biking around campus. To get more people on bikes, Environment America started a program for students that let them trade in their car parking permit for a $400 voucher at all local bike shops. So, the program increased the campus bike-riding rate by 100%.
Public transit	College campuses have the opportunity to leverage the existing web of public transport to create convenient, low-cost campus commutes. This one is consistent and predictable. In collaboration with the transportation providers, make comfortable bus stops to promote the use of public transportation on college campuses. It also helps to provide discounted or free public transit passes to students and employees. The ride privilege offers them the flexibility for planning commutes, while the campus gains a smaller carbon footprint and fewer cars on its property. As those parking lots go unused, campuses can also trim their budgets for maintaining them. Visual communication is the other factor that matters. Students may be provided with signs, directions, and orientation maps through the local area. Students also required apps that provide them up to date information about their timetables so that they can plan their day accordingly.

Electric vehicles	Electric vehicles (e.g., electric scooters) need infrastructure to be used, but renewable energy and sustainable communities are two of the sustainable development goals. Campus electrification initiatives are deploying charging stations, aided by state and local programs that support the initiatives and help to subsidize fleet electrification on campuses. In addition to it, universities can also plan to join hands with electric mobility players who have tie-ups with educational institutions. Some countries, like North American campuses, have operational e-scooters and e-bicycles with Bird. It meets the special needs of college communities for efficient and eco-friendly transit. Every shared mobility system, including private university bike rentals and public e-scooter rides, consists of two primary software components: the Dashboard and the Rider App. As the command centre, the dashboard shows a detailed picture of every linked car, battery, and docking station in real time. Administrators may track use trends, keep an eye on each vehicle's position, and oversee maintenance plans via this backend management system. In order to enforce parking regulations, slow-speed zones, and regions off-limits to e-bikes and e-scooters, geofence borders will also be defined on the Dashboard. Without any paperwork or manual procedures, students may access campus-owned automobiles using the Rider App. By using the app, students may find bikes and scooters in their area, scan the QR code on the vehicle, and quickly get to class or other local facilities. Other useful features offered by the Rider App include trip history, price information, safety instructions, and student help. These two software elements work together to provide a seamless environment for campus electric car management when combined with dependable, high-quality hardware.
Smart resources	Universities provide students with the resources they need to transition to eco-friendly transportation. Universities are creating flexible online programs that allow students to take control of their sustainability journey. These may be e-transportation platforms via which students place their opinions on important issues and interventions. Carpooling apps based on campus can connect them with others to travel with. Smart Commute Trip Planner by University of Maryland (United States) is a Successful Initiative It encourages students to find transportation and plan their journeys, and explore local transport, bike paths and walking paths. There are also gamification aspects to the platform. Students can log points that they have captured by making their commutes greener and redeem them in the Smart Commute store.

Source: own study.

As the world continues to grapple with climate change, universities have started to realize that green travel initiatives are a rigorous first step in minimizing their footprint on the planet. By transforming how campuses move, colleges can help to lower greenhouse gas emissions, promote exercise, alleviate congestion, and even make it cheaper to get to campus. Table 5.5 presented case studies of sustainable solutions for university mobility.

Table 5.5 Case studies of sustainable solutions for university mobility

University	*Program/ initiative*	*Description*
University of Malta (Malta)	Green Travel Plan (GTP)	Since 2010, the program has been running on campus. This has helped to make certain parts of campus parking more logical. Many awareness-raising events on using public transportation, cycling, or walking to university have been held as part of this GTP. Students and faculty may utilize an interactive web-based bike facilities map that shows the locations of bike racks, restrooms, drinking water fountains, and bike maintenance services. The primary focus for 2017 was raising awareness of pedestrian issues, which included starting a walking ambassador network and making maps of pedestrian routes from various communities.
University of Glasgow (Scotland)	Green Transport Policy	Since 2006, the program has been running with well-defined goals that align with the Sustainable Urban Mobility Plans (SUMPs) methodology. This SUMP aims to: identify, support, and promote alternatives to single occupancy car travel; improve access routes and amenities around university grounds for sustainable travel; and improve staff and student fitness and health by promoting walking and bicycling. Several travel plans have been used to accomplish this SUMP, the most recent of which being the "University of Glasgow Strategic Transport and Travel Plan 2016–2025."
The University of Bristol (UK)	Combined Staff and Student Travel Plan	The University of Bristol (UK) is spread across multiple buildings or faculties in the city of Bristol, the majority of which are located within the urban area. The university began implementing a green travel policy in 1998 when they began collecting travel data, and a significant part of their work has been the ongoing improvements in the methods of data collection and analysis. For instance, in 2007, they started weighing the samples based on whether travellers were full-time or part-time students or staff at the university. Their current "Combined Staff and Student Travel Plan" covers all locations and both staff and students, concentrating on the major transportation issues that impact the university's strategic goals.

(*Continued*)

Table 5.5 (Continued)

University	*Program/ initiative*	*Description*
The University of La Coruna (Spain)	Combined perspective of public space and mobility	There are two locations for the University of La Coruna in Spain: Coruna and Ferrol. This institution began a really well-thought-out strategy in 2010 that blended an intriguing viewpoint on mobility and public space. In the new mobility paradigm outlined in the plans, less private vehicle journeys would be made in favour of more environmentally friendly modes. The renovation of public areas, the removal of surface parking, and many parking spots are also included in the proposal. This strategy is very socially aware and uses a participatory approach, wherein the university community helped determine the overarching goals.
University of Milan (Italy)	The mobility management project	The goal of the University of Milan's mobility management project was to find, specify, and evaluate intervention strategies to lessen the environmental effects of university employees' and students' movement. The plan's results included developing and implementing mobility action plans across university campuses to optimize and regulate flows and provide a framework for addressing mobility management issues.
Roma Tre University (Italy)	Sustainable mobility plan	In partnership with the Municipality of Rome, Roma Tre University developed a sustainable mobility plan that reduced the use of private vehicles in favour of group transportation options. The plan included low-impact transportation options like bicycles or electric motorcycles, collective public transportation, and the creation of an application for carpooling and carsharing on campus.

Sources: own study based on Papantoniou et al. (2020) and Travelling Smart-Green Transport Plan for The Msida Campus, University of Malta (2011).

Internationally, a number of transportation strategies and regulations have been implemented to raise the standard of mobility on college campuses. One example is the technique used in Barcelona to connect the city transit plan with the university plan by implementing superblocks. Another noteworthy example is the University of Bristol, which has connected its several locations around the city by creating a joint staff and student travel schedule. The Polytechnic of Turin offers yet another significant example (Sgarra, Meta, Saporito, Persia, & Usami, 2022). Here, the plan suggested a system of control between bike lanes and parking, based on the mobility demands and transportation habits of all university members (Papantoniou et al., 2020). The Sapienza Sustainable University Mobility Plan (henceforth SUMP), in keeping with the most contemporary international universities conscious of pursuing eco-sustainable mobility policies, was born out of the necessity to understand and enhance, in the direction of sustainability, the modes of transportation for the staff and students of one of the oldest and largest universities in the world, with its premises situated in an urban context as complicated and full of challenges as the city of Rome. The goal of SUMP has been to develop strategies to maximize staff and student travel in a systematic manner with the goal of reducing the use of private vehicles in favour of eco-friendly forms of transportation and meeting staff mobility needs while adhering to the goals of lowering energy consumption, environmental.

5.3 Smart Traffic Management and Parking Solutions

Traffic management is vulnerable to the butterfly effect, which is a phenomenon in chaotic systems (urban traffic) where a little change in circumstances (one jaywalker) may have a significant impact on the whole system (town wide gridlock). It's not possible to have everyone follow traffic laws or capture all the butterflies. However, may set up a situation in which a single occurrence doesn't stop the traffic system as a whole. More eyes are needed in the community by urban planners to (Intellias, 2024):

- Get traffic situation information in real time.
- Identify issues and take prompt action.
- Implement preventative actions proactively.

According to Digi International, smart traffic management systems are technological solutions that cities can include into their junctions and traffic cabinets right now to quickly and affordably enhance traffic flow and safety on their city streets. According to Juniper Research, by 2027, smart traffic management technologies may reduce CO2 emissions by 205 million metric tons. It is obvious that intelligent solutions may significantly lower our carbon footprint, (2022). Next-generation hardware and software are used by effective and successful smart traffic management systems to maximize traffic infrastructure (Symmetry Electronics, 2025) (Table 5.6).

Smart traffic management systems, which use linked technology, provide a systematic, integrated method of reducing traffic and enhancing roadway safety on university campuses (Symmetry Electronics, 2025). According to PR Newswire, with a 10.1% CAGR, the market for intelligent traffic management systems is projected to reach $19.91 billion by 2028 (Research, 2021).

Smart technology is changing the way students navigate campus. On a university campus, smart technology can be used to monitor and manage pedestrian traffic on busy crosswalks, keeping students safe while minimizing disruptions to vehicle flow, especially during class changes. Using real-time data, technology-enabled traffic management systems can adjust street lights and redirect vehicles, reducing delays and increasing safety for everyone, especially in busy pedestrian areas. These systems constantly monitor traffic using

Table 5.6 Types of hardware and software to optimize traffic infrastructure

Hardware	*Software*
IoT road sensors including: • RFID (radio frequency identification), AIDC tags (automatic identification and data collection) • Temperature sensors	**Cloud computing and edge processing capabilities**: • Traffic data platform/data lake • Cloud-based traffic control system • Geographic information system (GIS)
Air quality sensors	**Transportation supporting applications**
Connected CCTV cameras	**Big data and predictive analytics**
Connected traffic light systems	**Artificial Intelligence/Machine Learning**: • Computer vision • Optical character recognition (OCR) • Reinforcement learning
Smart toll gates/electronic road pricing gantry systems	**Local based services**
Edge devices – chips on edge nodes for faster data processing	

Source: own study based: (Symmetry Electronics, 2025).

sensors and cameras, identifying bottlenecks and adjusting signals accordingly. For example, during peak hours, they can extend green lights on major roads to keep traffic moving, while during quieter times, they prioritize crosswalks and traffic. This flexibility not only reduces travel times, but also reduces accidents by managing traffic flow more efficiently. Additionally, this data can be used to inform how the institution approaches maintenance projects, providing insight into which routes are most used and when. By navigating these major thoroughfares or scheduling work during off-hours, facilities teams can minimize the impact on campus life during construction. These intelligent traffic systems also contribute to a greener environment by reducing downtime and unnecessary stops or bottlenecks. This can help reduce vehicle emissions, leading to cleaner air for the university and surrounding communities (Executive, 2024).

The issue of parking spaces has extended across urban growth. Because of this, the number of cars on campus is growing faster than the number of parking spaces accessible for both staff and students. The implementation of parking management for the smart campus may help address this problem by saving time, improving parking space usage, lowering management expenses, and reducing traffic congestion by matching vehicles to available parking spaces on an individual basis. An IoT Raspberry Pi-based parking management system (IoT-PiPMS) might be the example. Its purpose is to make it simple for employees and students to locate open parking spaces using GPS coordinates and real-time vision, all via a smartphone application. A Raspberry Pi 4 B+ (RPi) embedded computer, a Pi camera module, a GPS sensor, and ultrasonic sensors make up the system. The RPi 4 B+ is used in the IoT-PiPMS to collect and analyse data input from the sensors and camera. The data is then sent to the Blynk IoT server over Wi-Fi. To assure data accuracy, the Pi camera is used in conjunction with ultrasonic sensors and LEDs to identify parking place occupancy. By using the Blynk App, which finds parking spaces online, the GPS module is integrated into the system to help drivers find parking spots (Jabbar, Wei, Azmi, & Haironnazli, 2021).

Smart technology provides a revolutionary answer to the problem of parking, which is sometimes a significant cause of annoyance in metropolitan settings, particularly on college campuses where students are pressed for time to find a spot and go to class. It is no longer necessary to drive around the block in search of a parking place since sensors can identify open spots and route cars to them in real time. This considerably lessens the effect of wasted fuel on the environment and on-road congestion. Similar to how they are used at the municipal level, parking spot sensors may be connected to applications that monitor availability and provide staff and students real-time data. To avoid needless confusion, they may even be programmed to differentiate between parking categories, directing members of the community to the proper guest places. This function further reduces traffic and improves the general effectiveness of parking management on campuses and in cities (Executive, 2024). Table 5.7 shows case studies of universities with smart traffic and parking solutions.

Table 5.8 shows top 5 benefits and challenges of smart traffic management system.

Table 5.7 Case studies of universities with smart traffic and parking solutions

University	*Technology used*
University of California Berkeley (United States)	IoT sensors, AI-based traffic flow analysis, smart parking apps
Massachusetts Institute of Technology (United States)	IoT sensors, AI-driven traffic management
University of Sydney (Australia)	Smart parking, IoT-based congestion monitoring
University of Melbourne (Australia)	Smart parking systems, traffic flow optimization using IoT
University of Toronto (Canada)	IoT sensors, smart parking solutions, traffic flow management
Peking University (China)	Dynamic parking management, IoT sensors for traffic flow optimization
Zhejiang University (China)	AI-powered parking systems, real-time traffic monitoring with IoT sensors

Source: own study.

Table 5.8 Top 5 benefits and challenges of smart traffic management system

Benefits	*Challenges*
Improved traffic flow: Improved traffic management enables smoother, more efficient movement of vehicles, which means less congestion.	**Expensive implementation**: The initial configuration and continued maintenance of the systems require significant investments
Enhanced safety: The system reduces accidents by efficiently managing traffic signalization and flow	**Privacy and security**: Collecting traffic information risks violating the privacy of individuals and ensuring the security of collected data.
Cost reduction: Automating flow can help road networks operate more efficiently while reducing delays, resulting in significant savings in fuel and time for drivers and public transportation.	**Integration struggles**: Combining the fresh technologies with current foundations and systems may have posed obstacles, particularly in older universities with established arrangements.
Real-time monitoring: Improved emergency vehicle response times are made possible by reduced traffic, and in the case of a major public disaster (such as a tornado, earthquake, flood, etc.), smart traffic management systems' real-time insights enable better flexibility, action, and organization.	**Unreliable systems**: Dependence on detectors and advanced solutions could lead to matters if the frameworks faltered or if the data misrepresented.
Minimized emissions: More efficient use of road networks with reduced delays has the potential to yield substantial fuel and time savings for both private motorists and public transit systems.	**Ongoing tuning**: Guaranteeing that the system stayed effective necessitated continuous observing, refreshes, and alterations to accommodate changing conditions demanded continuing refinement.

Source: own study.

References

An, J. (2024). Route positioning system for campus shuttle bus service using a single camera. *Electronics*, *13*(11), 2004.

Autonomous campus mobility platform (2018). https://digital.wpi.edu/downloads/dr26z0144

Azzali, S., & Sabour, E. A. (2018a). A framework for improving sustainable mobility in higher education campuses: The case study of Qatar University. *Case Studies on Transport Policy*, *6*(4), 603–612.

Azzali, S., & Sabour, E. A. (2018b). A framework for improving sustainable mobility in higher education campuses: The case study of Qatar University. *Case Studies on Transport Policy*, *6*(4), 603–612. https://www.sciencedirect.com/science/article/pii/s2213624x18301342

Azzali, S., & Sabour, E. A. (2018c). A framework for improving sustainable mobility in higher education campuses: The case study of Qatar University. *Case Studies on Transport Policy*, 6(4), 603–612.

Balsas, C. J. (2003). Sustainable transportation planning on college campuses. *Transport Policy*, *10*, (1), 35–49. https://www.sciencedirect.com/science/article/pii/s0967070x02000288

Block, F., Keller, M. R., & Negoita, M. (2024). Revisiting the hidden developmental state. *Politics & Society*, *52*(2), 208–240.

Bond, A., & Steiner, R. (2011). Sustainable campus transportation through transit partnership and transportation demand management: A case study from the University of Florida. *Berkeley Planning Journal*, *19*(1), 125–142.

Campus automated shuttle service deployment initiative (2018). Tampa, FL: University of South Florida.

Car ownership index — Turo (2025). https://turo.com/ca/en/car-rental/canada/car-ownership-index [18 January 2025].

Clifford, J. M., & Cooper, C. D. (2012). A 2009 mobile source carbon dioxide emissions inventory for the University of Central Florida. *Journal of the Air & Waste Management Association* 1995, *62*(9), 1050–1060.

Climate Action (2025). *2030 Climate targets*. https://climate.ec.europa.eu/eu-action/climate-strategies-targets/2030-climate-targets_en [15 January 2025].

COAST (2025). COAST. https://www.coastautonomous.com/ [16 January 2025].

Ercan, T., Onat, N. C., Keya, N., Tatari, O., Eluru, N., & Kucukvar, M. (2022). Autonomous electric vehicles can reduce carbon emissions and air pollution in cities. *Transportation Research Part D: Transport and Environment*, *112*, 103472. https://www.sciencedirect.com/science/article/pii/s136192092200298x?casa_token=og_y4mgveu4aaaaa:wv6wuchtxnqi33v_chxgjfmn3hocztgfi52pldeoy5gavfhagvrtvz9ysn2ggqd7cr_vqnw

Erdoğan, S., Cirillo, C., & Tremblay, J.-M. (2015). Ridesharing as a green commute alternative: A campus case study. *International Journal of Sustainable Transportation*, *9*(5), 377–388.

Executive, F. (2024). *How smart cities can impact college campus life*. https://facilityexecutive.com/how-smart-cities-can-impact-college-campus-life/ [18 January 2025].

Exploring the role of attitude in the acceptance of self-driving shuttles (2020). https://escholarship.org/uc/item/1509n72d

Fu, X., Alabama, N., Liu, J., & Jones, S. (2022). *How do college students perceive future shared mobility with autonomous vehicles? A survey of the University of Alabama students*. Morressier GmbH.

Gava, G., & Willaschek, M. (2023). Transcendental doctrine of method. In *The Kantian Mind* (pp. 130–141). Routledge. https://www.taylorfrancis.com/chapters/edit/10.4324/9781003406617-16/transcendental-doctrine-method-gabriele-gava-marcus-willaschek

Hickman, R., Lopez, N., Cao, M., Lira, B. M., & Biona, J. B. M. (2018a). "I Drive outside of Peak Time to Avoid Traffic Jams—Public Transport Is Not Attractive Here." Challenging discourses on travel to the University Campus in Manila. *Sustainability*, *10*(5), 1462. https://www.mdpi.com/2071-1050/10/5/1462

Hickman, R., Lopez, N., Cao, M., Lira, B. M., & Biona, J. B. M. (2018b). "I Drive outside of Peak Time to Avoid Traffic Jams—Public Transport Is Not Attractive Here." Challenging discourses on travel to the University Campus in Manila. *Sustainability*, 10(5), 1462. https://www.mdpi.com/2071-1050/10/5/1462

How do college students perceive future shared mobility with autonomous vehicles? A survey of the University of Alabama students (2022). https://www.sciencedirect.com/science/article/pii/s2046043021000940

Iclodean, C., Cordos, N., & Varga, B. O. (2020). Autonomous shuttle bus for public transportation: A review *Energies*, *13*(11), 2917.

Intellias (2024). *Intelligent traffic management systems: A lowdown of software & hardware components*. https://intellias.com/intelligent-traffic-management/ [18 January 2025].

Jabbar, W. A., Wei, C. W., Azmi, N. A. A. M., & Haironnazli, N. A. (2021). An IoT Raspberry Pi-based parking management system for smart campus. *Internet of Things*, 14, 100387. https://www.sciencedirect.com/science/article/pii/s2542660521000317

Kim, S.-W., Gwon, G.-P., Hur, W.-S., Hyeon, D., Kim, D.-Y., Kim, S.-H., Kye, D.-K., Lee, S.-H., Lee, S., Shin, M.-O., & Seo, S.-W. (2017). Autonomous campus mobility services using driverless taxi. *IEEE Transactions on Intelligent Transportation Systems*, *18*(12), 3513–3526.

Latif, Y., Chowdhury, A., & Bagchi, S. (2024). *On-chain validation of Tracking Data Messages (TDM) using distributed deep learning on a Proof of Stake (PoS) blockchain*.

Lin, P.-S., Kourtellis, A., Menon, N., Chen, C., & Rangaswamy, R. & University of South Florida. Center for Urban Transportation Research (2018). *Campus automated shuttle service deployment initiative*. Tampa, FL: University of South Florida. https://rosap.ntl.bts.gov/view/dot/56111

Lin, P.-S., Menon, N., Chen, C., & Kourtellis, A. (2020). Assessment of a week-long campus automated shuttle demonstration in low-speed environments. In *Resilience and sustainable transportation systems* (pp. 263–272). Reston, VA: American Society of Civil Engineers.

Logan, K. G., Nelson, J. D., Osbeck, C., Chapman, J. D., & Hastings, A. (2020). The application of travel demand management initiatives within a university setting. *Case Studies on Transport Policy*, *8*(4), 1426–1439. https://www.sciencedirect.com/science/article/pii/s2213624x20301127

Medrano-Berumen, C., Malayjerdi, M., Akbas, M. I., Sell, R. & Razdan, R. (2020). Development of a validation regime for an autonomous campus shuttle. In *2020 SoutheastCon* (pp. 1–8). IEEE.

Nicefaro, M. (2019). Driverless shuttles as campus "people movers". *University Business*, 21 February. https://universitybusiness.com/driverless-shuttles-as-campus-people-movers/ [15 January 2025].

Ondruš, J., Kolla, E., Vertaľ, P., & Šarić, Ž. (2020). How do autonomous cars work?. *Transportation Research Procedia*, *44*, 226–233.

Papantoniou, P., Yannis, G., Vlahogianni, E., Attard, M., Regattieri, A., Piana, F., & Pilati, F. (2020). Developing a sustainable mobility action plan for university campuses. *Transportation Research Procedia*, *48*, 1908–1917. https://www.sciencedirect.com/science/article/pii/s2352146520306414

Pérez-Neira, D., Rodríguez-Fernández, M. P., & Hidalgo-González, C. (2020). The greenhouse gas mitigation potential of university commuting: A case study of the University of León (Spain). *Journal of Transport Geography*, *82*, 102550. https://www.sciencedirect.com/science/article/pii/s0966692319304764

Research, V. M. (2021). Intelligent traffic management system market size worth $ 19.91 billion, globally, by 2028 at 10.1% CAGR: Verified market Research®', *Cision PR Newswire* 25 November. https://www.prnewswire.com/news-releases/intelligent-traffic-management-system-market-size-worth--19-91-billion-globally-by-2028-at-10-1-cagr-verified-market-research-301432113.html [18 January 2025].

Saha, B., & Fatmi, M. R. (2021). Simulating the impacts of hybrid campus and autonomous electric vehicles as GHG mitigation strategies: A case study for a mid-size canadian post-secondary school. *Sustainability*, 13(22), 12501.

Salonen, A., & Haavisto, N. (2019). Towards autonomous transportation. passengers' experiences, perceptions and feelings in a driverless shuttle bus in Finland. *Sustainability*, *11*(3), 588. https://www.mdpi.com/2071-1050/11/3/588

Sgarra, V., Meta, E., Saporito, M. R., Persia, L., & Usami, D. S. (2022). Improving sustainable mobility in university campuses: The case study of Sapienza University. *Transportation Research Procedia*, *60*, 108–115. https://www.sciencedirect.com/science/article/pii/s2352146521009157

Smart traffic management systems to save 205 million metric tons of CO_2 by 2027 16 May2022. https://www.juniperresearch.com/press/smart-traffic-management-systems-to-save-205-million-metric-tons-of-co2-by-2027/ [18 January 2025].

Symmetry Electronics (2025). *What is a smart traffic management system?* https://www.symmetryelectronics.com/blog/what-is-a-smart-traffic-management-system/ [18 January 2025].

Toor, W., & Havlick, S. W. (2004). Transportation and sustainable campus communities: Issues, examples, solutions. *1559636564*. https://trid.trb.org/view/703008

Travelling smart-green transport plan for the Msida Campus, *University of Malta* (2011).

UBC Okanagan Transportation Plan (2025). https://campusplanning.ok.ubc.ca/planning-and-development/plans-and-policies/transportation-planning/ubc-okanagan-transportation-plan/ [15 January 2025].

US EPA (2015). *Fast facts on transportation greenhouse gas emissions | US EPA*. https://www.epa.gov/greenvehicles/fast-facts-transportation-greenhouse-gas-emissions [15 January 2025].

Waymo (2025). *Waymo - self-driving cars - autonomous vehicles - Ride-Hail*. https://waymo.com/ [16 January 2025].

Zhang, W., Guhathakurta, S., Fang, J., & Zhang, G. (2015). Exploring the impact of shared autonomous vehicles on urban parking demand: An agent-based simulation approach. *Sustainable Cities and Society*, 19, 34–45. https://www.sciencedirect.com/science/article/pii/s221067071530010x?casa_token=hdf-vsbjo6gaaaaa:zofucca66lg0jiabffqdbtjjrcubpjzgakuxqgcd2_cxnj_xxet2kwa5ufqsyl4vh1bptaw

2025，智能驾驶系统-百度智能驾驶解决方案及汽车智能化产品-百度Apollo|Baidu阿波罗. https://www.apollo.auto/apollo-self-driving [16 January 2025].

6 Data-Driven Decision-Making in Smart Universities

6.1 Big Data and Analytics for University Operations

Ranging from student demographics, enrolment trends, course performance, and faculty workloads, large volumes of data compiled via university operations yield so many valuable leads in furthering efficiencies and decision-making processes. This, facilitated by advanced analytics tools, enables a university to outline patterns and predict forthcoming needs that will aid efficient resource allocation and smooth running of operations (Jawwad, Turab, Al-Mahadin, & Owida, 2024).

Scheduling has been one of the big areas where big data and analytics have made a big strike. Traditional scheduling always degenerated into overcrowding of students in the classroom, scheduling conflicts, and underutilization of facilities. Today's institutions, driven by predictive analytics, will be able to forecast course demand in light of historical data and enrolment trends. Administrators are able to construct timetables that keep tabs on student preference while at the same time minimizing conflicts and optimizing the use of classrooms. In view of this fact, the algorithmic examination of the pattern in which students register can predict courses at peak demand, enabling ample sections to be provided to students without undue strain on resources (Yu, 2024).

The data-driven approach has revolutionized academic planning. Big data helps universities track the performances of students in real time and detect those who are at risk, providing further academic support accordingly. A study of grades, attendance, and engagement could be used to appropriate interventions that improve student success and retention. Analytics will help the university assess the quality and sufficiency of the program and determine which classes might experience problems requiring enhancements to the curriculum or further reinforcement to raise the level of achieved learning outcomes. Predictive models can estimate the number of students who will graduate and give a university an opportunity to identify in advance those specific areas of concern that keep students from getting their degrees (Liu, Tong, Mao, & Zhao, 2023; Mitrofanova, Aleksandrov, Ivanova, Nemtcev, & Popova, 2021).

Another area where big data and analytics bring enormous benefits is resource planning. Colleges and universities have to strike a very fine balance

DOI: 10.4324/9781003621829-7

across budgets, personnel, and facilities operations if they are to work effectively (Mitrofanova, Burenina, Tukshumskaya, Kuznetsov, & Popova, 2022; Sun, 2022; Wolniak, Gajdzik, Grebski, Danel, & Grebski, 2024; Wolniak & Stecuła, 2024a). Data analysis will allow administrators to view the trend in the use of resources and make decisions concerning staffing levels, equipment purchases, and building maintenance. Energy consumption patterns can be viewed for energy conservation, statistics about library use can be obtained for buying fresh material and redesigning study spaces. Workforce analytics provides the insight for higher education to analyse faculty workload and rebalance the load with fairness and productivity in mind. In Table 6.1, there is an analysis

Table 6.1 The applications of big data in smart university

Aspect	*Application of big data and analytics*	*Benefits*
Scheduling	Optimize your class scheduling and timetabling through right-on-point enrolment and course demand trends.	Minimizes scheduling conflicts, reduces overcrowding, improves classroom utilization, and enhances flexibility for students.
Academic planning	Monitor the performance to predict academic risk in determining the better effectiveness of programs.	Improves student retention, enhances learning outcomes, identifies at-risk students, and supports targeted interventions.
Resource allocation	Better plan and efficiently manage your resources by understanding facility utilization, staffing workloads, and energy consumption.	Ensures cost efficiency, balances faculty assignments, enhances infrastructure management, and reduces waste.
Student support services	Identifying behaviour patterns and performances leads to proper advising and career counselling.	Improves student engagement, supports timely interventions, and enhances satisfaction and graduation rates.
Financial management	Identify expenditure habits and budget information, which would facilitate financial planning, and the position to understand where one should anticipate expenses.	Enhances budgetary efficiency, reduces costs, and supports strategic financial decision-making.
Admissions and enrolment	Use predictive models on the application trends to further optimize the recruitment strategy	Increases enrolment rates, targets ideal student demographics, and improves marketing campaigns.
Research and development	Data in use identifies trends of research, funding opportunities, and possible collaboration.	Boosts research productivity, attracts funding, and enhances institutional reputation.
Facilities management	Track usage and maintenance facility data to help plan repairs, renovations, and expansions.	Improves campus infrastructure, reduces operational downtime, and supports sustainable practices.

Source: Author's own work.

of the applications of big data in smart university (Rico-Bautista, Maestre-Góngora, Guerrero, Sanchez-Velasquez, & Barrientos-Avendaño, 2021).

Most institutions are demanding now the gathering and analysis of a huge volume of data in respect to trends and patterns that drive decisions of a strategic nature, including improvement in operations (Table 6.2). This calls for systematic collection of data from diverse sources like student records, metrics related to performance by faculty, figures of enrolment, financial statements, and usage logs of resources (Yang, Chen, Liu, & Luo, 2022). Modern universities implement a high-integrated data management system in which information is centralized, accessible, and reliable. It is actually in the integration of these student information systems, learning management platforms, and institutional databases that universities create comprehensive datasets and insights on everything from academic to administrative functions (Jonek-Kowalska & Wolniak, 2023; Tang, 2021; Wolniak & Stecuła, 2024b).

Once the data are collected, better techniques of analyses such as predictive modelling, machine learning algorithms, and data visualization tools will have to be applied to draw meaning from them. Enrolment trends will be studied by universities in order to make predictions for demands over programs and courses so course offerings can be planned in advance, with faculty loads possible (Rico-Bautista, Medina-Cardenas, Areniz-Arevalo, Maestre-Gongora, & Guerrero, 2020). This would help institutions in having early signals on the struggles students go through in academics in time for intervention; as such, they will ensure better retention and overall success. Resource utilization analyses will assist administrators in achieving efficiency in the scheduling and usage of resources (Lei & Huang, 2021; Phanichsiti, Thamnita, & Thipphayasaeng, 2023).

Big data analytics supports decision-making at the strategic level through, for instance, insights related to evidence-based institutional performances regarding the assessment of impacts of academic programs, faculty productivity measures, and other financial expenditure resources to their effective utilization. These range from data-informed budgetary assessments that enable an institution to commit its resources to those areas likely to make a real difference, whether that's expanding high-demand programs or improving campus facilities. Additionally, university research analytics help pinpoint the budding areas of study with funding opportunities and are placed to make them more competitive in grants and partnerships (Chai, Chai, Wang, & Quan, 2023; Osuwa, Katende, & Osuwa, 2019).

Forecasting future needs and challenges also follows data-informed approaches. On the other side, universities analyse historical trends and demographic shifts to anticipate the makeup of their student population in some future time and to inform strategies related to the recruitment and management of enrolment. Predictive analytics, therefore, allows universities to plan and prepare for ups and downs in resource demand so that infrastructure and staffing levels are aligned with institutional goals. On the other hand, big data allows universities to offer value-added services in support for both student

Table 6.2 Methods academic institutions gather and examine huge datasets

Challenge	*Description*	*Implications and impact*
Data infrastructure limitations	Providing adequate infrastructure for scalability and security in storing, managing, and processing large amounts of data.	This involves high investment in hardware, cloud solutions, and software systems, which can be expensive and logistically problematic.
Integration of diverse data sources	Integrating data from disparate systems, including legacy systems, student information systems, and financial systems.	It adds complexity and might also create compatibility issues and possibly inconsistencies or data silos.
Data privacy and security concerns	Ensuring regulatory compliance with the General Data Protection Regulation, FERPA, etc., for sensitive student and institutional information.	There is an increased need for cybersecurity measures, which can bring legal and ethical questions concerning the use of data.
Shortage of skilled personnel	Hiring and maintaining a team of data scientists, analysts, and IT staff who are proficient in big data technologies.	Stronger difficulty in analytics and interpretation of data would further decrease the pace of adoption and quality of insights.
Cost of implementation	Software licensing costs for data storage, analytics, and maintenance.	Budgetarily constraining, especially for smaller institutions, it may delay or limit big data adoption in their entirety.
Data quality and accuracy	The data must be clean, consistent, and reliable; analytics and decision-making will ensue accordingly.	Poor quality of data may lead to misleading insights, thus affecting the reliability of predictions and strategic decisions.
Resistance to change	Also, overcoming the cultural hurdles and sceptic faculty and staff in light of adopting such data-driven practices	Delays the shift to data-driven operations and may need extensive training and awareness programs.
Scalability issues	There is scaling up the data system to support high-volume institutional needs.	May require frequent upgrading and modification of existing systems, adding to costs and administrative complexity.
Data governance and management	Clearly outline the roles and responsibilities, together with the policies for access to, usage, and accountability of data at institution-wide levels.	Ensures compliance and reliability but requires ongoing monitoring and enforcement to maintain standards.
Real-time processing demands	Development of systems for analysis of real-time data to enable real-time decision-making.	Demands advanced technologies and faster processing capabilities, which may strain resources and expertise.

Source: Author's own work.

advising and career choices, thus enabling students to make informed decisions about the development of their academic or professional life (Lin & Luo, 2021).

In Table 6.3, there is a description of selected case studies how data-driven strategies have significantly improved campus administration and resource.

Table 6.3 Case studies how data-driven strategies have significantly improved campus administration and resource

University	*Country*	*Description of case study*
Georgia State University	United States	Georgia State University built a high-end predictive analytics system to track a range of student performance indicators, from grades to attendance, even course participation. Aggregating the data across different sources, the university constructed an early alert system designed to flag students who were at risk of dropping out-or even just in danger of academic difficulties. In fact, that very system has given them personalized support, from proactive advising to tutoring and mentoring programs. This data-informed approach has been instrumental in closing equity gaps and boosting student retention rates.
University of Edinburgh	United Kingdom	The University of Edinburgh decided to opt for Learning Analytics Systems that would help in deducing individual levels of engagement and performance shown by their students. From the development of staff and advisor dashboards based on information provided from virtual learning environments and student records about progress made towards the attainment of an early warning to struggle, creation of customized learning plans with feedback on improvement is very important. This also brings about the identification of defects in curriculum design, therefore creating course changes that sensitively take note of and meet their students' needs.
University of Melbourne	Australia	The University of Melbourne applied data analytics to track research productivity, analysed funding trends, and found potential collaboration opportunities. This system allowed tracking of grant applications, publication rates, citation impacts, and thus performance measurements were enabled to make evidence-based decisions. The university was able to align its research priorities with funding priorities through insights provided by the data, enhanced partnership building, and increased grant success rates. By analysing the patterns in research output, administrators supported faculty development and improved research impact.

Sources: Author's own work on the basis of Akers (2023), Barshay & Aslanian (2019), Buyya (2022), How the University (2024), Innovation (2019), Knowles (2022), and Learning (2024).

6.2 Enhancing Student Experience through Data Insights

Analytics reports on on-campus life relate various extra-curricular activities regarding satisfaction with residence halls, down to events at the campus level. The colleges then develop and improve on the various programs and facilities to meet the students' interests using the recorded trends and patterns. For instance, predictive analytics can help in the detection of the areas of students' disengagement with the learning process to develop targeted interventions in the form of new club offerings, wellness programs, or food service changes. Data captured from campus security systems and incident reports are analysed for better decision-making about safety measures that will provide a safe environment where students can achieve academic and personal success. Another area of insight into data involves academic support, in which institutions make use of learning management systems in order to track attendance, assignment submission, and grades on examinations to find early possibilities of students being at academic risk (Yan, Ma, & Ru, 2024).

Thus, advisers can do some proactive outreach and provide tutorials, workshops, or study groups with a view towards filling in gaps in learning before it gets out of hand. The schools gather feedback from instructors and peers with the intent of continuously developing instructional methods that are attuned to students' current learning styles and preferences. It also studies use patterns of academic support centres for the purpose of resourcing support extended hours or to offer additional services at the hottest times of the year. Personalized learning has been the Holy Grail of modern education, and finally, insights from data have let colleges create experiences that precisely meet strengths, weaknesses, and aspirations for each and every student. Learning analytics platforms give granular insight to instructors about students engaging with course materials. On their part, instructors then can refine the content and assignments more suitably in an attempt to meet their students' needs. Adaptive learning technologies make a learning path individual to each student through continuous adjustments of course lessons in both difficulty and pace using real-time performance data for mastering concepts at students' speeds. Colleges also apply data concerning career development to provide ways to the students for the courses in mentorship programs and internships that offer practical skills essential in fields (Chango, Cerezo, & Romero, 2022; Lin & Luo, 2021; Stecuła & Wolniak, 2022).

Embed data insights within these key dimensions of the student experience, and colleges ensure that the supporting infrastructure will be in place to promote student success along multiple dimensions: well-being, academic success, and personal development. This would, in turn, translate into better retention rates, improved graduation outcomes, and a more adaptive learning environment to better prepare students for the challenges beyond college. Only those truly committed to the effective use of data will continue to evolve and change with these ever-shifting expectations from today's diverse, dynamic student body. Advanced algorithms and machine learning unleash a range of data insights colleges can leverage in building personalized experiences that ensure student and institutional success (Yu & Wang, 2024).

It will predict the performance within an academic setting, considering a number of performance metrics in terms of grades, attendance, participation, and engagement in course material. Such historical data analysis may provide predictive models based on trends in student performance in different subjects or environments, even their future academic performance. A prediction of this nature would, therefore, be an early warning for educators and administrators about which students are most likely to struggle-in, so timely interventions may be extended to provide much-needed support through additional tutoring, counselling, or more-focused academic advising. It will assist in letting insights from the data ensure that tutors change their teaching practices in regard to pedagogies that would stick with the learners and those actually in need of re-strategizing for effective learning outcomes (Chen, 2024).

After all, analytics for understanding students' needs are ways to show understanding-peculiar to needs, preferences, and challenges of students. It will enable analytics on learning management systems, student surveys, and indeed any other channel from which feedback is provided-and far better views, really, as to which students might need extra support or help. Analytics will define where exactly pupils struggle with a given subject or course, where curriculum revisions are to be expected, or what teaching methodologies may bring improved results. This would equally serve to expose deficiencies in the structures of support available to them, such as mental health, career counselling, and academic advising, on which colleges can more concretely adapt. With the system of tracking extra-curriculum activities, social events, or the use of campus facilities, universities can track with much attention those aspects of campus life which meant most to a student and therefore stand true in making a demand from his/her perspective (Wang, 2024).

Analytics would then be in a position to make the experience so personalized that it instils a feeling of community or belonging in the students for better engagement and retention. Analytics shall recommend course selection, clubs, mentorship programs, and professional development based on the academic progress and other dimensions of a student in tune with their interest in extracurricular activities. Adaptive learning is intelligent in character, availing data analytics for personalized content to students at matched levels of difficulty and pace for each learning style. This could be tapped into on an institutional level to issue targeted communications in line with updates, encouragements, and information emanating from relevant academic journeys, clubs, and associations one may subscribe to (Arora, Chakraborty, Bhatia, & Yadav, 2024; Feher, 2021).

In all these, the general aim is to bring out how data analytics would allow colleges and universities to go a long way beyond forecasting the academic success or particular needs of students and actually provide responsive and supportive learning environments, proactive behaviour, much better student engagement, and high academic results. This team helps encode data-informed strategies within institutions, making the services personalized hence foreseen long before challenges that might be insurmountable. This therefore remains different in driving success for students and advancing an institution (Kim, Jitpakdee, Praditsilp, & Issayeva, 2025; Masood, Zakaria, & Kamarudin, 2024).

In Table 6.4, there is description of difficulties with implementing big data solutions in academic institutions.

Table 6.4 Difficulties with implementing big data solutions in academic institutions

Area	*Application of data analytics*	*Impact*
Forecasting academic results	Data analytics examines historical performance data on grades, attendance, and participation in order to predict future academic outcomes. Predictive models are developed to assess potential challenges or successes.	Allows for the identification of at-risk students earlier to ensure timely interventions are made possible. Informs educators on how to adjust teaching methods based on projected needs.
Assessing student requirements	Through surveys, learning management systems, and feedback, colleges can highlight specific academic and personal issues that the students may face, such as subject difficulty, emotional well-being, and the use of campus resources.	Provides insight into the struggles of students academically, in support service needs, and in curricular areas that need improvement to intervene with targeted responses.
Customizing services for engagement	Data helps in informing student engagement with co-curricular activities, events on campus, and academic support services. Identifying trends in the behaviour of students enables an institution to build personalized course, club, and activity recommendations.	Engages students more by aligning campus offerings with student interests and preferences. Fosters a connected and engaged campus culture.
Personalized learning experiences	Adaptive learning systems modify difficulty, speed, and even content based on real-time performance data to meet individual learning needs and move students through material at their own pace.	Engages students to better learning outcomes with personalized, interactive educational content. It reduces dropout rates by providing early intervention with personalized support.
Improving retention rates	Data insights are put to use in tracking the level of engagement, academic performance, and usage of student support services. The early warning system triggers warnings for students showing signs of disengagement or academic struggle.	Students will feel supported and valued, and their likelihood of staying at school will increase.

(*Continued*)

Table 6.4 (Continued)

Area	*Application of data analytics*	*Impact*
Optimizing support services	Data insights are utilized regarding tracking the degree of involvement in studies, academic results, and access to available student support services. Accordingly, an early warning system turns on in cases when students show lack of interest or struggle to cope with academics.	Improves the efficiency and availability of crucial services. Ensures that students receive support at the most critical moments of their journey, bettering their whole experience.
Enhancing communication	Analytics on the interaction of the student with counselling, career services, tutoring, and other support systems enable pinpointing demand patterns and areas needing further resources. Segment the students into various groups based on their academic performances, extracurricular activities, and level of involvement. Personalize communications which are sent via email etc. based on the above information.	It encourages healthy communication with the students by informing them of all updates, reminders, and words of encouragement to keep them motivated.

Source: Author's own work.

By unleashing the power of data, community colleges and universities communicate more effectively, amplify mental health efforts, and build stronger campus communities. This ensures timely and relevant information on their academic progress, campus events, and support services is better understood and provided to the students easily. All three platforms gather data on the performance and activity of students; this therefore enables the various institutions to send out personal notifications, reminders, and alerts. For instance, it will remind a student when he is facing problems with deadlines or provide information to the student on how to manage his time. The communication channels can be based on students' needs, interests, or concerns so that their needs are well addressed in the right manner and at the right time. This way, it will enable the colleges to keep their students more engaged and informed by making use of the data hence decreasing the likelihood of students being disconnected or alienated (Chernova, Mustafina, Jayabalan, & Al-Jumeily, 2024).

It means that data analytics lie at the root of the colleges to better their insight into student mental health needs and hence develop strategic interventions. Institutions often keep a record of patterns related to student behaviour through periodic questionnaires, visitors to a counselling centre, and the students' performance in studies. For example, there is tracking of behaviours associated with failing grades or high absenteeism or isolation from campus life (Hidayat, Hidayah, Rusmana, Hariko, & Tririzky, 2024). Data from mental health apps, wellness programs, and peer support initiatives can position colleges to monitor trends in stress, anxiety, and other mental health challenges. It is through such data analysis that institutions are better placed in the distribution of resources to offer more counselling services, create campaigns for raising awareness about mental health, and organize stress-relieving programs in accord with the students' needs. Data-driven insights underpin the development of peer support networks and mental health initiatives in order to make services responsive and proactive (Surendran, Arulkumar, Aruna, Sangamithrai, & Thangadurai, 2024).

Colleges nurture much better campus communities through data by tracking student involvement in extracurriculars, social events, and academic clubs, providing a sense of where their students may be dislocated or underserved. With such a facility, they would be able to organize more representative events catering to the diverse nature of the students by evaluating the participation pattern in every event. Further, institutions make use of data to promote clubs and activities in concert with students' academic and personal goals that would make students feel a sense of belonging and more deeply engaged in campus life. Understanding student preferences and involvement trends allows colleges to build active, supportive, and connected communities that better meet the diverse needs of their students (Purnama Alamsyah, Morika, Siswanto, & Othman, 2024).

While colleges are working to improve in terms of communication and mental health support, data integration has truly fostered a community atmosphere that cares about well-being and community involvement. A data-centred approach such as this keeps students connected, supported, and valued in their education for the pursuit of higher learning; thus, it contributes to overall satisfaction with the college experience. With each refinement in the use of the data at institutions, they can become better equipped to meet emerging challenges their students are facing, building a resilient and thriving campus community (Razak, Seriandi, Mien, Lim, & Nathan, 2024).

6.3 Data Privacy and Security in Smart Campuses

Data privacy and security in smart campuses have only just now become one of the most critical concerns, with the recent integration of advanced technologies at educational institutions. Such campuses, which feature the use of IoT, cloud services, big data, and AI, are supposed to provide seamless connected experiences to students, staff, and faculty. It is precisely this interconnectivity that enables such innovations to have created considerable challenges in maintaining sensitive information. Such data in smart campuses include academic, health information, and behavioural patterns; all of which is prone to breach or misuse unless properly safeguarded. Among the challenges that go along with ensuring data privacy and security within a smart campus is the volume and variety of data being collected. The IoT devices deployed in these campuses-from smart cameras to wearables and environmental sensors-keep generating data that is most likely to include PII. Often, this is broadcast across multiple networks: campus-wide Wi-Fi to cloud-based solutions; and sometimes interception is done which makes information vulnerable. Thus, securing the data at all points of its transmission or storing without compromising privacy on such a ground becomes tougher (Ahtinen, Chowdhury, Ramírez Millan, Wu, & Menon, 2023; Fidan & Boza, 2024).

The complication in this issue increases with the inclusion of various technologies and platforms, each having its security protocols and standards, hence creating a gap in protection. Increased use of cloud computing in smart campuses brings additional risks. While cloud services give scalability and flexibility, they raise a host of concerns related to data sovereignty and control. It would therefore mean that very little, if any, control will be left for the institutions over the storage, processing, and sharing of their data once they have been handed over to a third-party cloud provider. This then sets various challenges towards compliance with the local data protection laws, since different jurisdictions have varying standards concerning data privacy. Most of the complex data security management regarding both on-campus systems and an externally hosted cloud environment leaves so many gaps that always remain liable to be used by cyber criminals (Katavić, Ivanović, & Papić, 2023a; Zhou & Liu, 2022).

Another layer of challenges comes with the ever-growing sophistication in the nature of cyberattacks. These might be through phishing, ransomware, or even DDoS attacks. Besides this, with such a huge variance in devices, it will be quite difficult to ensure that each and every endpoint is kept under observation and secured. IT administrators have an overwhelming job to prevent data breaches and protect personal information with confidentiality and integrity, amidst the increasing volume of data and complexity of maintaining security across a highly dynamic, interconnected campus environment (Tanate-Lazo & Cabonero, 2021).

Apart from the technological challenges that include maintaining data security, there are several other human factors due to which a smart campus faces a very tough time. The fact is that, given an increasingly mobile student and faculty population, coupled with the proliferation of personally owned laptops as well as other digitally-enabled devices and equipment, human error and/or carelessness can occur. Hence, sensitivity of information disclosure would possibly occur due to making uses of unsecured network transmission or password sharing or negligence by not performing software enhancements, for example. 'Most sophisticated technical solutions could compromise by individuals'. It therefore follows that training the users on best practice for data privacy and security is as important as any other technical measure (Clark & Driller, 2020).

Laws and regulations related to data privacy go a long way to regulate the many ways approaches towards private data security within the educational setting are expressed. These are those that have to make their practice comply with the complicated legal landscape brought about by different frameworks, such as the General Data Protection Regulation in Europe, the Family

Educational Rights and Privacy Act in the United States, among others, in various countries or internationally. In fact, such regulations demand not just security measures but full compliance with what personal data really is and how it must be treated. This has turned quite challenging in this smart campus environment where the data has to be collected from every point in real time, consisting of both identifiable and non-identifiable information (Myat & Soe, 2020).

This is another factor that complicates safeguarding private data on smart campuses, the trade-off between convenience and security (Table 6.5). Whereas a smart campus system has been devised to bring convenience and an improvement in the learning environment, it created a trade-off between ease of access and vulnerability. For example, any system that allows students to access learning materials, communicate with their cohort and faculty, and then track their progress online will have to be both usable and integrated into existing campus systems. Sometimes, it is true, the ease with which such systems are to be used compromises the level of security. The key is often a balancing act: keeping user data open yet safe (Radway, Quintanilla, Ludden, & Votipka, 2024; Toth, Abraham, & Zimmermann, 2024).

Data breach and unauthorized access in the area of data privacy and security in the smart campus are threats that are pretty serious and multidimensional, considering their scale of impacts on both individual and institutional levels. Data breach, a condition where unauthorized people have accessed sensitive personal information, may lead to the leakage of highly confidential data such as academic records, health information, and personal identifiers. All these could always happen within such a closely knitted atmosphere as a smart campus, wherein there is operational IoT devices, cloud systems, and campus-wide networks. Such data, if not well taken care of, there is a likelihood of it being hacked or subjected to any other phishing for malicious means and may also lead to severe financial losses, reputation, and other legal consequences for both the students and the institution itself. A breach could be defined as, but was not limited to, theft of academic records or PII data. Even misappropriation of intellectual property further exacerbates the damage done (Du & Aviles, 2024; Özer Şanal & Çiçek, 2024).

Wherever the case is unauthorized access to data, especially on a large scale, there is a high probability of misuse or exploitation. Students, faculty members, and employees in smart campuses can also fall prey to cyber-attacks due to weak spots in the campus network to gain data for malicious purposes that might include identity theft or fraud. Apart from that, unauthorized access might lead to the monitoring of movements, activities, or communication of individuals on campus – a further breakdown in the sense of privacy integral to a safe and trusting academic environment. Personal consequences for those whose information has been compromised through such breaches are not confined to financial loss. It is emotionally and psychologically perturbing to expose one's private information to the world without consent, as it affects an individual's perception of security and well-being (Dao, Vo, Pham, & Fukuda, 2023).

Table 6.5 Problems of safeguarding private data in smart universities

Difficulty	*Description*
Volume and variety of data	A smart university generates a huge amount of data from a vast number of IoT devices, sensors, cloud systems, and personal devices. This huge pool of data further adds to the complexity in securing sensitive information.
Data transmission risks	Information often flows over several networks, such as campus Wi-Fi or cloud storage, opening it to potential interception, unauthorized access, or even data breaches while in transit.
Cloud computing risks	This means that universities rely on cloud services and risk losing control over how their data is stored, processed, and passed on, especially to third-party providers who may not be adhering to all local laws on data protection.
Cybersecurity threats	Sophisticated cyber-attacks such as phishing, ransomware, and DDoS take advantage of the weak points in the security measures of the system and might compromise a lot of personal information.
End-point security	More smart phones, laptops, wearables mean more endpoints that could serve as entry points for the cybercriminal. Thus, securing each endpoint on the network becomes increasingly challenging.
Human error and negligence	Students, faculty, and staff may inadvertently expose data by using insecure networks, sharing passwords, or neglecting to apply software updates. Human error alone can significantly undermine security efforts.
Data privacy compliance	The ever-changing landscape of data privacy regulations, such as GDPR and FERPA, presents a challenge to compliance given the volume and sensitivity of personal information that universities handle across multiple jurisdictions.
Balancing convenience and security	At the same time, making these systems easy for students and employees alike may compromise their security. For example, easily accessible learning objects may expose sensitive information if not all safeguards are in place.
Integration of diverse technologies	This wide integration of technologies from different departments, such as IoT devices, cloud platforms, and campus apps, may introduce security gaps in case not all technologies adhere to the same standards or protocols of security.
Real-time data collection	The continuous collection and processing of real-time data arriving from different sources raise many challenges in security and assure that sensitive information is not accidentally revealed during its usage.
Data breach detection and response	The reason it is difficult to detect is that breaches are hard in highly dynamic and complex systems. Early detection is required so that action can be taken in time to minimize the effect; this becomes too late due to volume and access points in the case of large data volume.
Legacy systems and compatibility	Older campus systems generally are not designed to understand or cope with modern security protocols. The integration of such legacy systems with newer and more secure technologies can introduce vulnerabilities unless carefully managed.

Source: Author's own work.

It would be a fallacy to dismiss the ethical issues related to data collection and use in a smart campus when considering the moral impact of a smart campus. Universities have a moral obligation, being the custodians of personal information that should be collected, stored, and put into use within an ethical framework for the service of best interests of the subject of the data. There is an ethical question intrinsic in the gathering and storing of data, both in its use. On one hand, integrated use of data-driven systems has the potential to significantly enhance all aspects of the educational experience, from personalized learning opportunities through to more efficient campus operations. Huge collections and analyses of personal data do raise important questions regarding consent, privacy, and autonomy. Is anyone aware of what data are collected, let alone how it may be used and shared? The question gets even graver when advanced analytics and AI keep up a seriously growing pace of extracting insight from personal data-even without a person being conscious, let alone with the consent of a data subject (Katavić, Ivanović, & Papić, 2023b).

Similarly, there is an imbalance in the bargaining power of smart campus data collection, in which institutions or companies possessing masses of personal information in their databases are much stronger compared to the individual. It would be regarded as exploitation when data subjects' personal information is utilized in a manner not consented to, such as targeted marketing, profiling, or surveillance. This will eventually make the institution a contributor to the erosion of that trust, since one begins to ask many questions about the collection and subsequent use of his data. The line separating legitimate use to improve learning and intrusive monitoring becomes even more blurred with continuous data collection from many devices, apps, and sensors (Dirin, Oliver, & Laine, 2023).

The use of data to predict, for instance, student performance or keep track of behavioural trends does bring up moral issues in relation to violation of privacy and loss of autonomy. Let me be clear: these technologies may help identify students in academic or emotional need, but they often codify bias when either the design or implementation of those technologies is flawed. Predictive algorithms, for instance, risk reinforcing stereotypes or decisions based on incomplete or biased data in ways that further marginalize certain sets of students. It is herein that the ethical concern for designers and implementors of such systems would be to build these systems with fairness, in a transparent manner, where accountability for the use of data is considered within the smart campus (Rahman, Rodzi, Razali, Nazri, & Al-Sharqi, 2023; Zambas, Illarionova, Christou, & Dionysiou, 2023).

References

Ahtinen, A., Chowdhury, A., Ramírez Millan, V., Wu, C.-H., & Menon, G. (2023). Co-learning around social robots with school pupils and university students - focus on data privacy considerations. In *ACM international conference proceeding series* (pp. 115–123).

Akers, K. (2023). *Predictive analytics to improve student outcomes*. https://www.ed.gov/sites/ed/files/about/offices/list/ope/predictiveanalyticsslides.pdf (Accessed 28.12.2024).

Arora, A., Chakraborty, P., Bhatia, M. P. S., Yadav, S., & Singla, N. (2024). Smartphone data analysis for nomophobic behavior among university students during lockdown in India. In *Proceedings of the 14th international conference on cloud computing, data science and engineering, confluence 2024* (pp. 228–231).

Barshay, J., & Aslanian, S. (2019). *Colleges are using big data to track students in an effort to boost graduation rates, but it comes at a cost*. https://hechingerreport.org/predictive-analytics-boosting-college-graduation-rates-also-invade-privacy-and-reinforce-racial-inequities/ (Accessed 28.12.2024).

Buyya, R. (2022). *University of Melbourne deploys arm-based, distributed container framework for fog computing on Oracle Cloud Infrastructure*. https://www.oracle.com/customers/university-of-melbourne-case-study/ (Accessed 28.12.2024).

Chai, Z., Chai, Z., Wang, M., & Quan, G. (2023). How will the smart logistics service of universities based on big data be transformed and developed': Taking Jiangnan University as an example. In *Proceedings - 2023 5th international conference on computer science and technologies in education, CSTE 2023* (pp. 265–269).

Chango, W., Cerezo, R., & Romero, C. (2022). Looking for the best data fusion model in smart learning environments for detecting at risk university students. In *Proceedings of the 15th international conference on educational data mining, EDM 2022*.

Chen, M. (2024). The era value and mode innovation of civic education in colleges and universities under the background of big data technology. *Applied Mathematics and Nonlinear Sciences*, 9(1), 1–17.

Chernova, N., Mustafina, J., Jayabalan, M., & Al-Jumeily, D. (2024). Sustainable learning process: Assessing the effectiveness of teaching methodology by analyzing spatial and temporal properties of a student as a subject. *Sustainability*, *16*(11), 4540.

Clark, M. I., & Driller, M. W. (2020). University student's perceptions of self-tracking devices, data privacy, and sharing digital data for research purposes. *Journal for the Measurement of Physical Behaviour*, *3*(2), 128–134.

Dao, H., Vo, Q.-H., Pham, T.-H., & Fukuda, K. (2023). I never trust my university for this! Investigating student PII leakage at Vietnamese Universities. *IEICE Transactions on Information and Systems*, *E106* (12), 2048–2056.

Dirin, A., Oliver, I., & Laine, T. H. (2023). A security framework for increasing data and device integrity in internet of things systems. *Sensors*, *23*(17), 7532.

Du, Y., & Aviles, J. S. (2024). The realization of protecting students' privacy information system in university information publicity. *Proceedings of SPIE - The International Society for Optical Engineering*, *13272*, 1327229.

Feher, K. (2021). Expectation of smart mentality and citizen participation in technology-driven cities. *Smart Structures and Systems*, *27*(3), 435–445.

Fidan, Ü., & Boza, N. A. (2024). Individual privacy should be an institutional value: Socio-technical design of university data collection systems. In *Data-driven business intelligence systems for socio-technical organizations* (pp. 366–384).

Hidayat, H., Hidayah, N., Rusmana, N., Hariko, R., & Tririzky, R. (2024). The effect of using smart application on critical literacy of engineering education students. *International Journal of Information and Education Technology*, 14(6), 834–844.

How the University uses learning analytics (2024). https://information-services.ed.ac.uk/learning-technology/more/learning-analytics/how-the-university-uses-learning-analytics (Accessed 28.12.2024).

Innovation in higher education (2019). https://agb.org/wp-content/uploads/2019/01/case_study_innovation_georgia.pdf (Accessed 28.12.2024).

Jawwad, A. K. A., Turab, N., Al-Mahadin, G., Owida, H. A., & Al-Nabulsi. (2024). A perspective on smart universities as being downsized smart cities: A technological view of internet of thing and big data, *Indonesian Journal of Electrical Engineering and Computer Science*, 35(2), 1162–1170.

Jonek-Kowalska, I., & Wolniak, R. (2023). *Towards sustainability and a better quality of life?* London: Routledge.

Katavić, S. S., Ivanović, H., & Papić, A. (2023a). Digital privacy and data protection knowledge and skills of university students in Croatia: Preliminary findings. *Education for Information*, *39*(4), 493–515.

Katavić, S. S., Ivanović, H., & Papić, A. (2023b). Digital privacy and data protection knowledge and skills of university students in Croatia: Preliminary findings. *Education for Information*, *39*(4), 493–515.

Kim, L., Jitpakdee, R., Praditsilp, W., & Issayeva, G. (2025). Does technological innovation matter to smart classroom adoption? Implications of technology readiness and ease of use. *Journal of Open Innovation: Technology, Market, and Complexity*, *11*(1), 100448.

Knowles, C. (2022). *The University of Melbourne implements data ecosystem with Talend.* https://itbrief.com.au/story/the-university-of-melbourne-implements-data-ecosystem-with-talend (Accessed 28.12.2024).

Learning analytics projects (2024). https://information-services.ed.ac.uk/learning-technology/more/learning-analytics/learning-analytics-projects (Accessed 28.12.2024).

Lei, J., & Huang, G. (2021). Smart Management system of employment in universities based on big data collection and intelligent analysis. In *Proceedings of the 5th international conference on trends in electronics and informatics, ICOEI 2021* (pp. 1110–1113), 9453076.

Lin, J., & Luo, W. (2021). Big data analysis model of smart campus education based on university students' behavior analysis. *ACM international conference proceeding series* (pp. 74–77).

Liu, F., Tong, X., Mao, J., & Zhao, Y. (2023). Design and development of big data platform for smart university. In *ACM international conference proceeding series* (p. 26).

Masood, A., Zakaria, N., & Kamarudin, M. A. I. (2024). Empowering e-learning excellence: Unveiling the influence of outcome expectation, learning motivation, and self-efficacy in the Industrial Era 4.0 (IR 4.0). *Journal of Advanced Research in Applied Sciences and Engineering Technology*, *42*(1), 254–264.

Mitrofanova, Y. S., Aleksandrov, A. Y., Ivanova, O. A., Nemtcev, A. D., & Popova, T. N. (2021). Smart university: Development of analytical management system based on big data. *Smart Innovation, Systems and Technologies*, *240*, 373–382.

Mitrofanova, Y. S., Burenina, V. I., Tukshumskaya, A. V., Kuznetsov, A. K., & Popova, T. N. (2022). *Smart university: Digital development projects based on big data, smart innovation, systems and technologies*, 305 SIST (pp. 230–240).

Myat, S. M., & Soe, T. N. (2020). Preserving the privacy for university data using blockchain and attribute-based encryption. In *2020 IEEE conference on computer applications, ICCA 2020* (p. 9022852).

Osuwa, A. A., Katende, J. O., & Osuwa, A. A. (2019). Perception of smart campus big data analytics in university. In *Proceedings of the 3rd world conference on smart trends in systems, security and sustainability, WorldS4 2019* (pp. 48–58), 8903933.

Özer Şanal, S., & Çiçek, B. (2024). University students' digital spaces in online learning communities: Implications for understanding, protecting and maintaining privacy. *British Journal of Educational Technology*, *55*(2), 654–667.

Phanichsiti, S., Thamnita, O., & Thipphayasaeng, P. (2023). Internet of things and big data for a sustainable smart university. *Journal of Higher Education Theory and Practice*, *23*(2), 117–128.

Purnama Alamsyah, D., Morika, D., Siswanto, B., & Othman, N. A. (2024). Smart campus and education: The study of individual performance in e-learning environments. In *11th international conference on ICT for smart society: Integrating data and artificial intelligence for a resilient and sustainable future living, ICISS 2024 – proceeding*.

Radway, S., Quintanilla, K., Ludden, C., & Votipka, D. (2024). An investigation of US Universities' implementation of FERPA student directory policies and student privacy preferences. In *Conference on human factors in computing systems - proceedings* (p. 791).

Rahman, A. B. A., Rodzi, Z. M., Razali, I. N. B., Nazri, I. S. B. M., & Al-Sharqi, F. (2023). Breaking the illusion: The reality of Artificial Intelligence's (AI) negative influence on university students. In *2023 4th International conference on artificial intelligence and data sciences: discovering technological advancement in artificial intelligence and data science, AiDAS 2023 - proceedings* (pp. 194–199).

Razak, N. A., Seriandi, G., Mien, M., Lim, L. Z., & Nathan, V. A. L. (2024). Moderating students' attitudes between communication, creation and information skills with perceived digital competency. *Pakistan Journal of Life and Social Sciences*, 22(1), 6348–6363.

Rico-Bautista, D., Maestre-Góngora, G. P., Guerrero, C. D., Sanchez-Velasquez, M. C., & Barrientos-Avendaño, E. (2021). Smart university: Key factors for the adoption of internet of things and big data. *RISTI - Revista Iberica de Sistemas e Tecnologias de Informacao, 2021*(41), 63–79.

Rico-Bautista, D., Medina-Cardenas, Y., Areniz-Arevalo, Y., Maestre-Gongora, G., & Guerrero, C. D.(2020). Smart university: Big data adoption model, applications in software engineering. In *Proceedings of the 9th international conference on software process improvement, CIMPS 2020* (pp. 52–60), 9390151.

Stecuła, K., & Wolniak, R. (2022). Advantages and disadvantages of e-learning innovations during COVID-19 pandemic in higher education in Poland. *Journal of Open Innovation: Technology, Market, and Complexity*, *8*(3), 159.

Sun, W. (2022). Construction of smart campus in universities under the background of big data intelligence. *Lecture Notes in Electrical Engineering*, *791*, 537–544.

Surendran, D., Arulkumar, V., Aruna, M., Sangamithrai, K., & Thangadurai, N. (2024). Improving the quality of education through data analytics and big data contributions. *AIP Conference Proceedings*, *2742*(1), 020001.

Tanate-Lazo, R. J., & Cabonero, D. A. (2021). Philippine data privacy law: Is it implemented in a private university library, or not? *Library Philosophy and Practice*, *2021*, 1–26.

Tang, Z. (2021). Analysis of information security problems and countermeasures in big data management of colleges and universities under smart campus environment. In *Proceedings - 2021 2nd international conference on information science and education, ICISE-IE 2021* (pp. 912–915).

Toth, A., Abraham, Z., & Zimmermann, V. (2024). Understanding university students' concerns regarding automated learning assessment tools. In *UMAP 2024 - adjunct proceedings of the 32nd ACM conference on user modeling, adaptation and personalization* (pp. 195–200).

Wang, Y. (2024). Resource integration of university students' ideological and political schooling website based on data mining technique. In *Lecture notes of the institute for computer sciences, social-informatics and telecommunications engineering, LNICST, 580 LNICST* (pp. 3–11).

Wolniak, R., Gajdzik, B., Grebski, M., Danel, R., & Grebski, W. W. (2024). Business models used in smart cities—Theoretical approach with examples of smart cities. *Smart Cities*, *7*, 1626–1669.

Wolniak, R., & Stecuła, K. (2024a). Evaluation of quality of innovative e-learning in higher education: An insight from Poland. *Applied System Innovation*, *7*(6), 109. https://doi.org/10.3390/asi7060109

Wolniak, R., & Stecuła, K. (2024b). Artificial intelligence in smart cities—applications, barriers and future directions: A review. *Smart Cities*, 7, 1346–1389.

Yan, H., Ma, Y., & Ru, Z. (2024). Computer-aided medical analysis of smart university student's psychological fitness data using fuzzy clustering and decision tree algorithms. *Computer-Aided Design and Applications*, *21*(S9), 277–291.

Yang, J., Chen, Y., Liu, X., & Luo, X. (2022). Design and practice of university smart campus integration platform under the background of big data. In *Proceedings - 2022 3rd international conference on information science and education, ICISE-IE 2022* (pp. 86–89).

Yu, H., & Wang, J. (2024). Construction of a big data-based student performance assessment and personalised instruction system in physical education teaching. In *2024 9th international symposium on computer and information processing technology, ISCIPT 2024* (pp. 167–171).
Yu, X. (2024). Research on smart contract application system integrating big data platform and blockchain technology in universities. In *2024 IEEE 2nd international conference on sensors, electronics and computer engineering, ICSECE 2024* (pp. 1198–1202).
Zambas, M., Illarionova, A., Christou, N., & Dionysiou, I. (2023). Exploring user attitude towards personal data privacy and data privacy economy. In *Lecture Notes in Networks and Systems, 721 LNNS* (pp. 237–244).
Zhou, Z., & Liu, Y. (2022). Blockchain-based encryption method for internal and external health privacy data of university physical education class. *Journal of Environmental and Public Health, 2022*, 7506894.

7 Enhancing Campus Life through Smart Services

7.1 Smart Healthcare Services on Campus

The smart healthcare services on university campuses stand for welfare; hence, offering innovative ways such that the use of advanced technologies including telemedicine, health monitoring devices, and AI-driven diagnostic tools can be offered. In the face of the growing trend of digitalization in every sphere of life, universities these days are increasingly inclined to adopt smart healthcare systems that efficiently facilitate access and personalization in care services to the increasingly diverse populations of their students.

From this perspective, one of the big revolutions has been telemedicine, enabling students to consult a doctor and receive health advice without necessarily attending the consultation. This is important in a university environment since students normally face constraints such as time, geographical distance, and even a reluctance to visit a healthcare facility. With telemedicine, it bridges these gaps by creating avenues for virtual consultation with healthcare professionals. This is through video calls, consultations by phone, or messaging, whereby students get expert medical advice, prescription, or referral without leaving their dormitory or classroom. Such a service does not only increase accessibility but also eases the pressure on university health centres because they can pay more attention to more complex cases with the assurance that the student population has timely access to primary care (Chao, Cheng, Liu, & Hsu, 2023; Lotfi, Rodić, Labus, & Bogdanović, 2024).

Wearable devices, together with health monitoring devices, really changed the approach of universities in managing students' health (Table 7.1). Inventions such as smartwatches and fitness trackers, even special medical sensors, are fast turning out to be tools for monitoring different aspects of the students' health. These kinds of devices offer real-time vital sign data on heart rate, sleeping patterns, and physical activities – even up to blood oxygen saturation levels. These are life-saving gadgets for chronic conditions among students, particularly those suffering from diabetes and asthma, through the delivery of alerts in cases of abnormal readings to both the student and healthcare providers. In such a way, it will be trying to identify, in a proactive way, all the possible issues before they turn out to be serious health concerns; thus, it allows timely

DOI: 10.4324/9781003621829-8

Table 7.1 Tools used in smart healthcare services in smart universities

Tool	*Description*
Telemedicine	This virtual access is the platform for obtaining healthcare services remotely by the students through video calls, calling, and messaging with health professionals. In such a way, students can conveniently and efficiently seek medical advice and treatment without actually visiting campus health centres.
Health monitoring devices	Smartwatches and fitness trackers are becoming popular wearables, which individuals strap on their bodies to collect real-time health data, from a person's pulse rate to sleeping patterns, physical activity, and other important signs. The monitoring is continuous. An alert is sent whenever something abnormal is detected, so that timely intervention can take place against the health hazard.
AI-driven diagnostic tools	AI systems analyse health data, medical history, and diagnostic tests to help health professionals diagnose conditions. The technology looks at symptoms, medical imagery, and even behavioural data in search of patterns indicative of the early beginnings of everything from mental health issues to infection and chronic diseases, thus allowing for quicker and more accurate diagnosis.
Mental health monitoring tools	AI-powered tools monitor students' behaviour, survey responses, and other signals that may spell anxiety, depression, or stress, among others. The tools are designed to provide early alerts for intervention that could link students with counsellors or mental health professionals.
Health data analytics	Aggregators or analytics platforms aggregate data from several health tools to identify the key student health trends for both mental health challenges and chronic conditions, or wellness needs. It is this kind of data that shapes the on-campus wellness programs and initiatives, focusing on such particular health issues among the student communities.
E-prescriptions	This would mean that health professionals would prescribe medication electronically through the provided electronic platforms to the students. This is because this model averts the use of paper prescriptions and hence facilitates the access of prescribed medication through local pharmacies or online delivery services to students.
Remote patient monitoring	This would be a continuous operation system whereby sensors and connected devices monitor patient health that can be remotely observed by healthcare professionals. It could be useful for students with chronic conditions whereby their state of health is monitored habitually, with changes dealt with promptly.
Digital health records	Such records are easily accessible and can be updated by healthcare providers in order to make efficiency and continuity of care better, taking privacy and security into consideration. The electronic system manages and maintains health information on the history, medication, and test results of the students only.
Virtual health consultations	This will be a place where health professionals are available for consultation, health screening, and follow-up through online services. In this way, it would ensure that students seek medical advice without necessarily having to visit a health centre, which for most busy students is not accessible or convenient.

(*Continued*)

Table 7.1 (Continued)

Tool	*Description*
Health education platforms	Digital health education platforms, resources on wellness, and self-care tips that respond to their needs; some platform arms students with healthy behaviours, prevention, and mental health through making informed decisions about health and well-being.

Source: Author's own work.

intervention. This information might also be taken up by the universities in relation to their health systems, and general student health could be recorded while looking at trends that would inform initiatives or wellness programming (Nagowah, Ben Sta, & Gobin-Rahimbux, 2023).

Artificially intelligent diagnosis tools will also form part of the smart healthcare ecosystem on campus. For example, artificial intelligence will apply to vast data analyses regarding symptoms, past conditions, and test results for insights that can be given to healthcare providers to aid them in making correct diagnoses on time. AI-driven applications will ensure early detection of a suite of conditions, from mental health to infections and chronic diseases, within the campus. For example, AI algorithms can analyse questionnaires or behavioural data about students for early signs of depression or anxiety and recommend counsellors or psychologists for referral. Other AI systems interpret medical images, such as X-rays or MRI scans, to help doctors make diagnoses more quickly and more accurately. Such tools will enable doctors to provide more personalized and accurate treatment plans, improving the patient outcome (El Kheir, Alnufaili, Alsaffar, Assad, & Alkhalifah, 2022).

It integrates these technologies into campus healthcare infrastructure and does more than just raise the quality of care by instilling attitudes towards a holistic approach to student well-being. Finally, combining telemedicine with wearable health devices and AI diagnostic tools will let universities create a more efficient, evidence-based, and approachable health experience. This also goes in tune with the approach to health currently in development-that is, prevention rather than treatment. With smart healthcare services, there is also a reduction in a lot of stigma that usually accompanies approaching medical services. They would be more willing to opt for telemedicine services or even wear wearables for health monitoring, knowing full well their privacy is guaranteed and that help is given out discreetly on their own terms (Jonek-Kowalska & Wolniak, 2023; Jonek-Kowalska & Wolniak, 2025; Katyal, Narula, Govindarajan, & Sahota, 2022).

The data would be collected and then used in an effort to improve public health at the university campus. In such a way, when the aggregated health data are analysed, universities will be conscious of the trends on mental health, substance use, or stress levels of students. It may also help drive targeted wellness programming, mental health resources, and community-building activities to meet their needs. The continuous analysis of data will be helpful in understanding what works for the universities best to make necessary decisions on investing in resources to support students' health (Kawamoto, Kukhareva, Weir, Brooke, & Del Fiol, 2021).

Real-time access to healthcare provides enormous benefits for the patient, shifting the paradigm of how medical services are both provided and received. This is very important in terms of ease with which health issues may be addressed without much waiting. Many healthcare delivery systems normally make the patients wait from the time the symptoms develop until they can finally see the doctor, resulting in worsening conditions, anxiety, and complications that are otherwise preventable. Real-time access through telemedicine, wearables, or remote monitoring offers avenues for advice, diagnosis, and even treatments to be obtained by the patients themselves, instead of waiting for an appointment and then traveling to facilities. This can be most critical at times when timely action holds the key to healthcare, as in emergencies or other urgent cases where even slight delays may lead to severe consequences (Kilinc, 2025).

Medical services, facilitated by technology, become efficient and thus improve significantly in access and efficiency. Telemedicine also facilitates real-time communication of the patient to health professionals without physically having to visit a health facility that may be overcrowded. Consultations by doctors via screen for diagnosis of symptoms, prescriptions, and recommendations of treatment are possible without necessarily requiring the patient to leave home. That saves a great deal of time but more importantly, opens healthcare services to people who may otherwise have barriers to care such as distance or issues of mobility, or maybe lack of transportation. Another benefit of telemedicine in treating chronic conditions may result in more regular follow-ups because patients do not require repeated office visits; treatment and monitoring continue via video linkage (Bhambri, Soni, & Tran, 2024; Elango, Manjunath, Prasad, Ramachandran, & Selvaraju, 2023; Poonia & Al-Alshaikh, 2024).

The use of health monitoring devices that continuously track vital signs and other health metrics further enhances real-time healthcare. These include the simple fitness trackers, medically sophisticated equipment capable of delivering a great deal of real-time data to the healthcare provider for appropriate decisions on time. Examples are wearable devices that can track heart rates, blood pressure, or glucose levels. These send notifications immediately to both the patient and their doctor in case of abnormal readings. A stream of health information allows early intervention because small health issues do not have the time to develop into bigger ones (Huanca, Torres, Rodriguez-Fernandez, & Núñez, 2024; Pandey, Ashwinbalaji, Prakash, & Giridharan, 2024). Clinical professionals can also establish the trends of changes in conditions, monitor patients remotely, as well as correct the treatment plans, which include adherence to prescriptive therapies with fewer intensive clinic visits. AI use has smoothened healthcare service provision by making diagnoses very fast and accurate. It can also analyse huge volumes of data in real time to inform healthcare providers about patterns that can help drive their decision-making processes based on up-to-date information. It may include examples such as

AI-driven diagnostic tools that support doctors in interpreting medical images, reading lab results, or even forecasting the conditions a patient might be experiencing. AI automates the routine and provides decision support to the healthcare professionals so that they can concentrate on providing personalized care, which helps in improving speed and quality of treatment. All such technological supports reduce human error, optimize resources, and improve the overall efficiency in healthcare delivery (Purushotham & Prabu, 2024; Ravikishan, Reddy, & Rao, 2024).

Technology makes coordination of care by all professionals in the circle of care much easier. Whereas in the traditional system, the record and treatment plans are fragmented among various providers, thereby ensuring continuity of care becomes challenging. But digital health records and interconnected systems have made sure that the healthcare provider gets to access the complete medical history of the patients anytime, hence enabling better collaboration. This will ensure that the care is coordinated, all professionals working with the patient are in concert, which minimizes the chance of miscommunication and reduces the necessity for redundant testing or procedures. Moreover, patients will be able to have more insight into and manage their own health because many digital platforms allow the patients to view their medical records, communicate with their providers, and track their progress, therefore encouraging a more patient-centred approach (Zha, Wang, Ma, Kumar, & Min, 2024).

Real-time access to healthcare encourages a paradigm shift toward prevention. The technological capability for continuous monitoring and timely intervention helps in identifying health risks before they become critical. Example: Many kinds of wearables can track early signs of heart diseases, diabetes, or any other chronic condition, thus enabling one to make lifestyle changes or receive treatments early that could prevent the escalation of those conditions (Katal, 2024; Routray, Ghosh, Li, & Rabie, 2024). The added advantage of this proactive approach is that it reduces the overall burden on healthcare systems; prevention of serious health conditions is far less costly and resource-intensive than treatment of advanced stages of disease. Technology integrated into real-time healthcare ensures much greater accessibility for underserved populations. In areas where access to health facilities may be less available, rural populations benefit quite significantly from telemedicine and mobile health technologies. People in very faraway places from the nearest hospital or clinic can consult virtually with a healthcare provider. There will no longer be wasted time on transportation to get there, or troubles regarding how to manage, when facilities are unavailable nearby. Besides, technology may also allow such people with some kind of disability or with limited access to conventional healthcare settings an opportunity for equal opportunity to preserve health (Adibi, Rajabifard, Shojaei, & Wickramasinghe, 2024; Majumder, Gochhait, & Paliwal, 2024) (Table 7.2).

In Table 7.3, there is a description of selected case studies connected with the usage of healthcare technology in academic campus.

Table 7.2 Method used to integrate campus health with larger metropolitan health

Integration approach	*Description*
Telemedicine collaboration	Allows students to access specialized healthcare services from metropolitan health networks remotely. Virtual consultations will allow students to connect with doctors and specialists in larger healthcare facilities without having to leave campus, which is convenient and expands access to care.
Referral systems to specialists	Campus health centres can establish referral systems with metropolitan hospitals or clinics that ensure students needing specialized care are smoothly transferred to the appropriate experts. This will ensure comprehensive treatment for conditions beyond the capacity of on-campus services.
Electronic health record integration	The unified health record system would share student health data from campus health centres to metropolitan networks for the continuity of care, a reduction in duplicative testing, and a full view of the medical history of students for informed decision-making.
Mental healthcare access	This will ensure that all students have access to the required mental health facilities with integrated metropolitan mental health facilities. In cases of severe or long-term conditions, students can be referred for psychiatric treatment, inpatient care, and therapy sessions.
Emergency care coordination	Campus health services ensure that students are referred promptly to metropolitan hospitals with emergency health facilities for serious and complicated medical conditions, making certain that quality healthcare interventions are provided in good time in critical health events.
Chronic disease management	Such regular care and follow-up can be especially effective in the campus health centres among students with chronic conditions, while easily referring those that may require specialized treatments or more advanced diagnostic testing and/or long-term management.
Health data analytics for prevention	By integrating data from campus health centres and metropolitan networks, surveillance of the health status of student populations can be enhanced. Trends and patterns may be identified, enabling prevention programs and wellness initiatives tailored to emerging health concerns proactively.
Shared resource networks	By linking up with metropolitan health networks, for example, universities can take advantage of advanced medical equipment, laboratories, and treatments without having to maintain all that on campus. In effect, this shared-resource model ensures that students have state-of-the-art healthcare access whenever they need it.
Insurance coverage integration	Coordinated insurance between campus health services and metropolitan health networks is an efficient way to receive medical care. The integration of insurance and billing will enable the university to lower the administrative burden for seeking care in the student population.
Collaboration on public health initiatives	The campus health centre and the metropolitan network also have a place in planning large-scale public health projects that could include vaccination campaigns, disease prevention programs, health education about campus-specific health needs and general community health concerns.

Source: Author's own work.

Table 7.3 Case studies of institutions that have effectively used healthcare technology

University	*Country*	*Description*
University of California, San Francisco (UCSF)	United States	The bold steps in the integration of digital technologies taken by UCSF into the Nursing curricula are state-of-the-art courses on informatics, telehealth, and analytics, thus better equipping the student for contemporary health-care settings. Students practically applied learning in various research projects demonstrating such technology applications. Besides, UCSF has programs designed for the exploration of the use of artificial and machine learning in patient care and diagnosis to ensure that students are prepared to address current healthcare challenges.
University of Washington School of Pharmacy	United States	The University of Washington School of Pharmacy, in collaboration with The Everett Clinic, received a grant to assess the impact of a computerized prescription system in outpatient settings. The goal is to improve medication safety and patient outcomes through information technologies in healthcare.
University of São Paulo	Brazil	The University of São Paulo has been in the vanguard of modernization and management of healthcare, with the adoption of large-scale information systems at its university hospitals. Such systems are designed to standardize patient care practices, optimize administrative processes, and improve overall service efficiency. The use of electronic health records by the university increased data management and access, thus enabling the ability to make diagnoses faster and more accurately. Besides, the university is into predictive analytics and artificial intelligence in monitoring health trends to prevent the outbreak of diseases. The programs also focus on training health professionals in the application of these technologies in clinical practice.

Source: Author's own work on the basis of Bouabida, Lebouché, & Pomey (2022), Brangnan (2024), Devine et al. (2021), Fox (2024), Health (2024), Mussi et al. (2023), Mussi, Pereira, Lacerda, & Santos (2018), Stoumpos, Kitsios, & Talias (2023), UWSOP, (2024), Washington's (2024).

7.2 Digital Engagement Platforms and Student Well-being

With technology taking over the face of the modern age, more and more digital tools have been put into use within universities and colleges in developing environments that give equal importance to personal growth alongside academic success. Such places have become contact zones of importance where students can engage and connect with each other, thus accessing resources important for well-being and belonging. The ultimate goal of all digital platforms would be the potential creation for reaching out and networking beyond face-to-face physical bounds. Many times, students have to struggle through academics, extra-curricular activities, apart from personal problems; hence, not everyone can always be physically present at any location or particular activity. It fills in by providing a digital channel wherein access to peers, faculty, and other support services from the comfort of a personal house or dorm is accessible easily. Such virtual places nurture a sense of belonging, and thus, discussions about joining virtual clubs, societies, and campus events also make students feel included in spite of the distances. The anonymity afforded by the digital platform lowers the threshold to sharing thoughts and feelings in areas where sharing might otherwise be unlikely-for sensitive areas related to help-seeking for, say, poor mental health (Elsayary & Olowoselu, 2024; Shao, Wu, Xu, & Zhang, 2024; Wolniak & Stecuła, 2024‘ Wolniak, Gajdzik, Grebski, Danel, & Grebski, 2024).

Such would play a variety of roles in advancing student well-being. In this respect, the social networking features, discussion forums, and peer support groups provide opportunities to establish networks for the purpose of social and emotional support. Such networks create a feeling of companionship, reducing the feeling of isolation common at university, and especially among first-year students and remotely studying students. Most of these integrated digital tools have features inbuilt that support the students in accessing well-being resources, counselling services, stress management programs, or even mindfulness exercises to make it easier for them to take a step forward with regard to their mental health (Brown, Lord, & John, 2023).

Digital platforms will afford an environment where access to information on student services and campus events could be so provided that no student, because of personal circumstance and schedule, gets left behind in terms of engaging opportunities and personal growth. It empowers campus communities by knowing that there is information to make an informed participation-be it attending a workshop virtually, hanging out in virtual social areas, or getting academic supports for success. Many of these digital spaces provide opportunities for students to connect with campus support staff in more agile and adaptive ways and give students the choice to efficiently and swiftly overcome obstacles. Faculty members and counsellors alike are guaranteed timely feedback or guidance through the digital platforms put in place for reassuring students in the right place both academically and personally. Such brings into perspective an important impact which encompasses valuing of diversity across all walks in life created through use of digital platforms (Büssing, Zupanic, Ehlers, & Taetz-Harrer, 2022; Joseph-Richard & Uhomoibhi, 2024). A system like this would allow for the easy outing of information, say, events dealing with culture, initiatives that have to do with the diversity of the institution's population, or campus social movements. It has the added advantage of allowing international and domestic students to create conversations that may not otherwise have taken place in physical surroundings. For instance, through the Internet, students can represent diversities, share lives, and show solidarities with others across cultures and identities that further reinforce inclusiveness on campus. Digital interaction platforms ensure the academic and social life of students remains in continuum during any period of crisis or disruption – for example, the COVID-19 pandemic. Digital platforms such as these have been a lifeline in connecting students to classes and extracurricular activities when physical interactions were not possible. It is this adaptability that has managed to retain the sense of community, with a number of lessons learned from how to build resilient and tech-savvy environments which will sustain student well-being when unexpected challenges strike (Hasan, AlMajaly, Alkhatatbeh, & Agha, 2024; Stecuła & Wolniak, 2022).

With these, important tools have emerged in building a dynamic and inclusive campus culture that goes beyond the classroom. Therefore, it would mean that effective integration of technology in campus life allows the organization to reach out to their students meaningfully and afford them more opportunities to get involved and be supported to grow towards total wellness. For example, club, sports, and cultural participation often required one to be physically present at a place and time. Indeed, digital spaces have created much room for flexibility in terms of how students participate in or even organize such activities. They are able to join virtual clubs, attend workshops online, and even participate in webinars or virtual events in subjects of interest to them without any geographical or time constraint. That serves several users, especially when one has an intensive academic schedule, is a remote learner, or is committed to something personal. Maybe most importantly, the digital platforms provide that hub through which students explore the wide range of extracurricular offerings, so as to ensure for every student that there is a community with interests like theirs out there (Ahammed & Smith, 2019; Pham, Phan, & Do, 2023).

These virtual networks establish and nurture peer support networks that mean so much for students' emotional and social well-being. Life as a college student is by no means easy, with scores of problems relating to academic pressures, homesickness, and personal struggles. With online forums, chat groups, and virtual meeting places made accessible through the use of technology, students have ways of sharing experiences, expressing themselves, and mutually supporting one another. Such networks provide a sense of belonging important to mitigations of feelings of isolation and cultivate a nurturing atmosphere. A lot of platforms allow students to create study groups, connect over majors, or create and join an affinity group based on shared identities and experiences. A network of peers becomes an outlet where students could discuss challenges, share successes, and advise each other in a safe and understanding place (Noor, Younas, Saleh Aldayel, Menhas, & Qingyu, 2022). Where for some students, mental health is increasingly a big concern, affected by anxiety, depression, or even stress related to school, social challenges, and even personal levels. This access to mental health facilities through digital platforms has gotten much easier and more 'invisible' for students to begin with. In return, this would allow students to avail counselling services via virtual sessions, attend workshops on mindfulness, and make use of the self-help resources while being in the comfort of their own space in privacy. A number of such portals ensures that mental health support services can be availed 24*7 to help students anytime, irrespective of geographical boundaries or time zones. It empowers the universities in making their students aware of messages pertaining to mental health awareness, coping mechanisms, and forthcoming events regarding issues on mental health through digital platforms so as to make campus culture welcoming and supportive (Abdelmagid, Hafez, Ahmed, Teleb, & Aljawarneh, 2024; Hu, Yu, Tzeng, & Zhong, 2023).

7.3 Smart Security and Safety Solutions

Installing newer gadgets at various universities speaks volumes in bringing down the quotient of vulnerability within the campus and thereby offering betterment to security apparatus. The system spoke beyond mere traditional physical security personnel or even surveillance cameras; rather, advanced technologies were integrated within the same, ensuring real-time monitoring and threat detection for prompt responses. Another important benefit of smart security solutions is the possibility to provide comprehensive, connected systems that allow continuous surveillance and monitoring across diverse campus areas. Smart cameras, with the potential of AI, will scan footage in real time and point to unusual behaviour or signs of a potential threat related to unauthorized access to some places, congregation of groups of people, or some kind of aggression. Being a real-time collection of data, an incident will automatically trigger the warnings to security personnel for effective actuation and response in record times, thereby enhancing general safety (Majeed, Majdi, Mohammed, Hassooni, & Almaamori, 2024).

Besides, the access control systems may allow smart cards, biometrics, and even mobile applications to get inside the building or rooms securely, for which smart security solutions can be provided. Such systems can also be designed for specific access permissions so that a distinct group of people could enter certain places, say dorms, laboratories, or even administrative offices. It finally obstructs plenty of misfeasance incidents such as unauthorized intrusions to particular areas or rooms, finally keeping both the students and staff members safe (Oughannou, Atmani, Kandrouch, el Houda Chaoui, & Chaoui, 2022).

An increasing number of campuses have initiated personal safety apps, most of which feature an emergency line direct to the security personnel or campus police. Many of these include GPS tracking where the security teams are able to pinpoint whence a distress call has been made. Besides, some include virtual escort features, services whereby the students can be tracked moving using their smartphones from and to classes and late-night activities. This will give more confidence to the students; hence, it will increase proactive means for personal safety since students can be assured help will come at the right time (Huang, 2021).

Some other major applications of the environment sensor include the following apart from the smart security implementation. It is necessary these days to install sensors over quality of air along with fire detections and alarm systems on most of the campuses. In case of any leakage of gas or fire within the campus, automatic notifications are provided, integrated in-depth into a networked system to provide timely alerts routing to the relevant authorities for minimal damage in case of an emergency (Tang, 2021).

Advanced security implementations make use of real-time monitoring. These systems record events in an environment in real time and can give information, on a minute-to-minute basis, of the status of physical space from the behaviour of occupants in an area. From surveillance cameras and access controls to environmental sensors, this all ensures through real-time monitoring that everything out of place-from unauthorized entry and suspicious activity to safety hazards-is immediately picked up and flagged for response. For example, monitoring real-time security threats allows for prompt interventions mostly before the situations get out of hand, which is a plus in comparison to the other conventional ways of security. Further, routing through a central control or even mobile phones ensures that the security personnel are connected at all times and are prompt when duty calls. With the changing roles of IoT devices, the definition of security systems has been redesigned (Ramadan, Al-Daweri, & Al Muniri, 2024). The general understanding of the IoT is basically a network of things interconnected, doing communications amongst them using the internet. In the context of security, IoT devices like smart cameras, motion detectors, smart locks-even wearable devices-create a dynamic, interactive environment that one can monitor and control remotely (Assiri, 2023). A good example could be that the smart camera will not only shoot but also analyse video information in real time for possible threats by comparing movement patterns or using face identification. Such locks can be fitted at all access points leading to restricted areas in a way that only authorized people enter into sensitive areas and send alarms to security teams in case of any breach. These devices continuously feed data into centralized systems, building streams of information that help in making better decisions faster from the observed anomalies (Ahmad, Zhu, Li, Kumar, & Chen, 2024; Musaev, Rakhmatullaev, Normatov, Shukurov, & Abdullaeva, 2024).

This AI, whether it is a real-time monitoring system or IoT devices, will go a long way in handling volumes of data with great efficiency. In this case, the algorithms developed by the AI will analyse and compare this output from sensors and cameras through trend identification, forecasting, or independent decision-making across multiple situations. For instance, AI-powered surveillance cameras analyse crowd behaviour, flag suspected activities involving aggressive movements or trespassing for theft by sending notifications to security personnel. Also, predictive analytics uses AI to forecast where an incident may occur, using historical data and patterns that help the security team manage resources and take preventive measures accordingly. That is to say, through time, the AI gains continuous learning and adaptation, hence making those systems increasingly accurate and precise; therefore, improving security in the environment with increased time (Hernández-León & Rodríguez-Conde, 2024). All put together, real-time monitoring systems, IoT devices, and AI integrate into a network for seamless operations in maintaining a secure environment. This will integrate a highly proactive security infrastructure wherein early detection of threats would not only be feasible but automated or initiated with minimum latency. These technologies contribute to larger efficiency, reducing human interventions and thereby letting the security personnel devote time and skills to critical tasks while the routine monitoring and threat identifications autonomously come via the system. More importantly, it allows the flow within the system to continue undisturbed so that contextual awareness about the security landscape may be possible in real time, with appropriate execution of emergency response measures (Santos, Claro, Medrado Gondim, & Mane, 2024) (Table 7.4).

From providing restrictive, automated access to smart camera surveillance and networked, emergency response systems, which are more efficient and safer to students, faculty, and other staff, campus security keeps developing as an effective strategy for improvement. Advanced technologies complement each other in highly interconnected and responsive security infrastructures that can take steps proactively to prevent the occurrence of incidents and will quickly address arising threats with speed (Elomiya, Krupka, & Jovcic, 2023).

Table 7.4 The use of technology in security

Technology	*Description*	*How it contributes to security*
Real-time monitoring systems	Systems that track and observe activities or conditions continuously in real time.	It can find security threats that take immediate effect, such as intrusion or suspicious activities. It allows for fast intervention and decision-making. Coordinated command and monitoring centres and control rooms for different security points. Smart cameras, motion detectors, sensors provide real-time data in view of security analysis. Smart locks and the access control system limit entry to an authorized person.
IoT devices	Interconnected devices that communicate over the internet, sharing data and performing actions.	Wearables and other devices send notifications to a security team in an emergency or track location. AI enhances the processing of data from monitoring systems and IoT devices, including the detection of behavioural anomalies or intrusion attempts.
Artificial intelligence (AI)	Algorithms that analyse and interpret data from various sources to make decisions or predictions.	It can automate responses, everything from locking doors to alerting authorities. It also introduces predictive analytics, predicting where a security incident is most likely to occur and optimizing resources accordingly.

Source: Author's own work.

The access systems are a great deal more secure and smooth with the incorporation of technologies like biometric identification, smart cards, or even mobile applications. Such systems can be installed to allow access to specific areas at only particular times of the day, for instance, depending on an individual's authorization or any other predefined set of criteria. For example, access may be restricted in academic buildings at certain hours of the day, while dorms require a bit more security access controls in respect of protection of occupants' safety and privacy. The automated access systems reduce unauthorized entry, hence improving campus security with real-time records of who comes in and out of particular areas, enabling campus security teams to monitor and track movements in and around sensitive spaces (Al Ghazo & Hayajneh, 2023; Almeida, 2023).

The smart surveillance cameras enhance the security through real-time monitoring of the campus environment. Unlike CCTV systems, smart cameras are designed with advanced technologies like AI and face recognition that enable them to analyse footage by themselves and detect unusual behaviour. These cameras are installed to detect people staying in places where they should not be, spot a person in an area that is supposed to be free of any people, or even fight and vandalize. Once there is something suspicious, the system alerts the security personnel who may investigate further. This allows for the detection of security risks well in advance, quicker responses to incidents so that no incident gets out of proportion. Besides, many cameras with cloud-based storage securely archive footage that might later be reviewed for incident investigation or in dispute resolution (Balint, 2021; Matos & Santos, 2023).

The ethical implications from such technologies go deep. Facial recognition is one good example, especially since debate has come to the front on the technology's accuracy and the probable emergence of discriminatory practices. Generally, large datasets form the basis on which algorithms making up AI are usually trained, but these might be incomplete and biased. This might give unjustified identification or lack of accurate ones, influencing different demographic groups differently. This raises ethical concerns in misidentifications and propagation of systemic biases in larger systems of security. Additionally, the lack of transparency-a general trend in data collection, use, and dissemination-results in little control over an individual's private information, therefore undermining any confidence in the institutions that implement those technologies (Abar, Letaifa, & El Asmi, 2022).

References

Abar, T., Letaifa, A. B., & El Asmi, S. (2022). QOE-based intelligent intrusion detection Use case: University video surveillance. In *2022 IEEE 9th international conference on sciences of electronics, technologies of information and telecommunications, SETIT 2022* (pp. 7–11).

Abdelmagid, A. S., Hafez, M. A., Ahmed, E. W., Teleb, A. A., & Aljawarneh, N. M. (2024). Interactive digital platforms and artificial intelligence applications to develop technological innovation skills among saudi university students. *International Journal of Interactive Mobile Technologies*, *18*(11), 64–79.

Adibi, S., Rajabifard, A., Shojaei, D., & Wickramasinghe, N. (2024). Enhancing healthcare through sensor-enabled digital twins in smart environments: A comprehensive analysis. *Sensors*, *24*(9), 2793.

Ahammed, F., & Smith, E. (2019). Prediction of students' performances using course analytics data: A case of water engineering course at the university of south Australia. *Education Sciences*, *9*(3), 245.

Ahmad, I., Zhu, M., Li, G., Kumar, P., &Chen, S. (2024). A secure and interpretable ai for smart healthcare system: A case study on epilepsy diagnosis using EEG signals. *IEEE Journal of Biomedical and Health Informatics*, *28*(6), 3236–3247.

Al Ghazo, A. T., & Hayajneh, M. R. (2023). Advanced IoT-AI security system with drone surveillance: Campus smart security prototype. *International Journal on Engineering Applications*, *11*(5), 346–357.

Almeida, F. (2023). Prospects of cybersecurity in smart cities. *Future Internet*, *15*(9), 285.

Assiri, B. (2023). Blockchain technology framework for universities. In *10th International conference on ICT for smart society, ICISS 2023 – proceeding*.

Balint, K. (2021). The connection of a blockchain with students' attendance register based on security cameras. In *SISY 2021 - IEEE 19th international symposium on intelligent systems and informatics, proceedings* (pp. 67–70).

Bhambri, P., Soni, R., & Tran, T. A. (2024). *Smart healthcare systems: AI and IoT perspectives* (pp. 1–346). Boca Raton: CRC Press.

Bouabida, K., Lebouché, B., & Pomey, M. P. (2022). Telehealth and COVID-19 pandemic: An overview of the telehealth use, advantages, challenges, and opportunities during Covid-19 pandemic. *Healthcare (Basel)*, *10*(11), 2293.

Brangnan, L. (2024). *Telehealth*. https://healthinformatics.ucsf.edu/telehealth (Accessed 2.01.2025).

Brown, M., Lord, E., & John, A. (2023). Adaptation of ACTivate your wellbeing, a digital health and well-being program for young persons: Co-design approach, *JMIR Formative Research*, *7*, e39913.

Büssing, A., Zupanic, M., Ehlers, J. P., & Taetz-Harrer, A. (2022). Mental stress in medical students during the pandemic and their relation to digital and hybrid semester—cross-sectional data from three recruitment waves in Germany, *International Journal of Environmental Research and Public Health*, *19*(17), 11098.

Chao, P.-J., Cheng, Y.-H., Liu, Y.-H., Hsu, T.-H. (2023). University students' attitudes toward adoption of mobile health application for serving older people. *Sensors and Materials*, *35*(7), 2581–2601.

Devine, E. B., Hansen, R. N., Wilson-Norton, J. L., Lawless, N. M., Fisk, A. W., Blough, D. K., Martin, D. P., & Sullivan, S. D. (2021). The impact of computerized provider order entry on medication errors in a multispecialty group practice. *Journal of the American Medical Informatics Association*, *17*(1), 78–84.

El Kheir, D. Y. M., Alnufaili, S. S., Alsaffar, R. M., Assad, M. A., & Alkhalifah, Z. Z. (2022). Physicians' perspective of telemedicine regulating guidelines and ethical aspects: A Saudi experience. *International Journal of Telemedicine and Applications*, *2022*, 5068998.

Elango, S., Manjunath, L., Prasad, D., Ramachandran, G., & Selvaraju, S. (2023). Super artificial intelligence medical healthcare services and smart wearable system based on IoT for remote health monitoring. In *Proceedings - 5th international conference on smart systems and inventive technology, ICSSIT 2023* (pp. 1180–1186).

Elomiya, A., Krupka, J., & Jovcic, S. (2023). A smart parking system using surveillance cameras and fuzzy logic: A case study at Pardubice University's campus. *Procedia Computer Science*, *225*, 4881–4890.

Elsayary, A., & Olowoselu, A. (2024). *Interdisciplinary approaches for educators' and learners' well-being: Transforming education for sustainable development* (pp. 1–271). New York: Springer.

Fox, A. (2024). *UCSF taps health informatics leader to head new digital transformation division*. https://www.healthcareitnews.com/news/ucsf-taps-health-informatics-leader-head-new-digital-transformation-division (Accessed 2.01.2025).

Hasan, M., AlMajaly, S., Alkhatatbeh, R., & Agha, N. (2024). Digital transformation revolution and higher education universities in Jordan: Exploring the students' attitude toward the use of eLearning platforms in the COVID-19 pandemic. *Studies in Systems, Decision and Control*, *489*, 243–252.

Health information (2024). https://healthinformatics.ucsf.edu/ (Accessed 2.01.2025).

Hernández-León, N., & Rodríguez-Conde, M.-J. (2024). Artificial intelligence applied to education and educational assessment at the university: Introduction of intelligent tutoring systems, recognition systems, and other future trends. *Revista de Educación a Distancia*, *24*(78), 6.

Hu, Y.-H., Yu, H.-Y., Tzeng, J.-W., & Zhong, K.-C. (2023). Using an avatar-based digital collaboration platform to foster ethical education for university students. *Computers and Education*, *196*, 104728.

Huanca, F., Torres, M., Rodriguez-Fernandez, M., & Núñez, F. (2024). A cyber-physical system for real-time physiological data monitoring and analysis. *IEEE Internet of Things Journal*, *11*(17), 28918–28930.

Huang, Z. (2021). An analysis of information security and protection strategies in colleges and universities in the environment of smart campuses. *Journal of Physics: Conference Series*, *1852*(4), 042050.

Jonek-Kowalska, I., & Wolniak, R. (2023). *Towards sustainability and a better quality of life?* London: Routledge.

Jonek-Kowalska, I., & Wolniak, R. (2025). *Exclusion in smart cities Assessment and strategy for strengthening resilience*. London: Routledge.

Joseph-Richard, P., & Uhomoibhi, J. (2024). Which data sets are preferred by university students in learning analytics dashboards? A situated learning theory perspective. *INFORMS Transactions on Education*, *24*(3), 220–237.

Katal, N. (2024). AI-driven healthcare services and infrastructure in smart cities. In *Smart cities: Blockchain, AI, and advanced computing* (pp. 150–170).

Katyal, N., Narula, N., Govindarajan, R., & Sahota, P. (2022). Setting Up a teleneurology clinic during COVID-19 pandemic: Experience from an academic practice. *International Journal of Telemedicine and Applications*, *2022*, 4776328.

Kawamoto, K., Kukhareva, P. V., Weir, C., Brooke, B. S., & Del Fiol, G. (2021). Establishing a multidisciplinary initiative for interoperable electronic health record innovations at an academic medical center. *JAMIA Open*, *4*(3), ooab041.

Kilinc, H. H. (2025). An encryption scheme and performance results for IoT-based healthcare service framework. In *Lecture notes in networks and systems, 977 LNNS* (pp. 347–3540.

Lotfi, F., Rodić, B., Labus, A., & Bogdanović, Z. (2024). Smart healthcare: developing a pattern to predict the stress and anxiety among university students using machine learning technology. *Journal of Universal Computer Science*, *30*(10), 1316–1342.

Majeed, A. S., Majdi, H. S., Mohammed, H. A., Hassooni, A. S., & Almaamori, M. H. (2024). Design and implementation of iot smart system for sensing toxic gas emissions and security system of chemical laboratories or stores of universities. In *2024 10th international conference on automation, robotics, and applications, ICARA 2024* (pp. 557–564).

Majumder, K., Gochhait, S., & Paliwal, M. (2024). Bibliometric analysis of Internet of Things (IoT) in green healthcare. *Lecture Notes in Electrical Engineering*, *1194*, 619–629.

Matos, H., Santos, H. (2023). People tracking in a smart campus context using multiple cameras. In *CEUR workshop proceedings*.

Musaev, M., Rakhmatullaev, M., Normatov, S., Shukurov, K., & Abdullaeva, M. (2024). Integrated intelligent system for scientific and educational information retrieval. *Vide. Tehnologija. Resursi - Environment, Technology, Resources*, 2, 212–219.

Mussi, C. C., Luz, R., Damázio, D. D. R., Santos, E. M. D., Sun, V., Porto, B. S. D. S., Parma, G. O. C., Cordioli, L. A., Birch, R. S., & Andrade Guerra, J. B. S. O. D. (2023). The Large-scale implementation of a health information system in Brazilian University Hospitals: Process and outcomes. *International Journal of Environmental Research and Public Health*, *20*(21), 6971.

Mussi, C. C., Pereira, C. D., Lacerda, R. T. O., & Santos, E. M. (2018). Pre-Implementation evaluation of a nationwide information system for university hospitals: Lessons learned from a study in Brazil. *Behaviour & Information Technology*, *37*, 217–231.

Nagowah, S. D., Ben Sta, H., Gobin-Rahimbux, B. A. (2023). Semantic model for IoT-enabled smart healthcare - an ontology-based study. In *6th International conference on inventive computation technologies, ICICT 2023 - proceedings* (pp. 1297–1302).

Noor, U., Younas, M., Saleh Aldayel, H., Menhas, R., & Qingyu, X. (2022). Learning behavior, digital platforms for learning and its impact on university student's motivations and knowledge development. *Frontiers in Psychology*, *13*, 933974.

Oughannou, Z., Atmani, A., Kandrouch, I., el Houda Chaoui, N., & Chaoui, H. (2022). IoT security in smart university systems. *Studies in Big Data*, *111*, 247–268.

Pandey, K. V., Ashwinbalaji, P. M., Prakash, R., & Giridharan, N. (2024). Secure patient healthcare monitoring system with IoT and body sensor network. In *Proceedings - 3rd international conference on advances in computing, communication and applied informatics, ACCAI 2024*.

Pham, D. V., Phan, V. C., Do, N. H. (2023). A new approach for visual analytics applying to multivariate data of student intakes in the university. In *Lecture notes of the institute for computer sciences, social-informatics and telecommunications engineering, LNICST, 473 LNICST* (pp. 88–120).

Poonia, R. C., & Al-Alshaikh, H. A. (2024). Ensemble approach of transfer learning and vision transformer leveraging explainable AI for disease diagnosis: An advancement towards smart healthcare 5.0. *Computers in Biology and Medicine*, *179*, 108874.

Purushotham, S., & Prabu, S. (2024). *Machine learning and generative AI in smart healthcare* (pp. 1–443). New York: IGI Global.

Ramadan, R. A., Al-Daweri, M. S., & Al Muniri, I. S. (2024). Data mining for enhanced security: A transformative framework for smart grid protection. *Journal of Engineering Research*, *21*(1), 23–32.

Ravikishan, S., Reddy, B. E., & Rao, K. V. S. (2024). An optimised task scheduling of remote sensing data processing for smart patient health monitoring. *Remote Sensing in Earth Systems Sciences*, 7(4), 542–553.

Routray, S., Ghosh, U., Li, X., & Rabie, K. (2024). Editorial AI driven internet of medical things for smart healthcare applications: Challenges and future trends. *IEEE Journal of Biomedical and Health Informatics*, *28*(6), 3279–3281.

Santos, C. N. M., Claro, D. B., Medrado Gondim, J., Mane, B. (2024). Suspicious behavior detection near vehicles in university environment: An approach using object detection and body angles. In *ACM international conference proceeding series* (p. 66).

Shao, Y., Wu, J., Xu, W., & Zhang, C. (2024). The impact of digital technology use on adolescents' subjective well-being: The serial mediating role of flow and learning engagement. *Medicine (United States)*, *103*(43), e40123.

Stecuła, K., & Wolniak, R. (2022). Advantages and disadvantages of e-learning innovations during COVID-19 pandemic in higher education in Poland. *Journal of Open Innovation: Technology, Market, and Complexity*, *8*(3), 159.

Stoumpos, A. I., Kitsios, F., & Talias, M. A. (2023). Digital transformation in healthcare: Technology acceptance and its applications. *International Journal of Environmental Research and Public Health, 20*(4), 3407.

Tang, Z. (2021). Analysis of information security problems and countermeasures in big data management of colleges and universities under smart campus environment. In *Proceedings - 2021 2nd international conference on information science and education, ICISE-IE 2021* (pp. 912–915).

UWSOP brings health technology into the 21st century (2024). https://sop.washington.edu/uwsop-brings-health-technology-21st-century/ (Accessed 2.01.2025).

Washington's Everett clinic: E-prescribing in a fast-changing health IT environment (2024). https://digital.ahrq.gov/ahrq-funded-projects/past-initiatives/transforming-healthcare-quality-through-health-it/washingtons-everett-clinic-e (Accessed 2.01.2025).

Wolniak, R., & Stecuła, K. (2024). Evaluation of quality of innovative e-learning in higher education: An insight from Poland. *Applied System Innovation, 7*(6), 109. https://doi.org/10.3390/asi7060109

Wolniak, R., Gajdzik, B., Grebski, M., Danel, R., & Grebski, W. W. (2024). Business models used in smart cities— theoretical approach with examples of smart cities. *Smart Cities*, 7, 1626–1669.

Zha, B., Wang, Z., Ma, L., Kumar, S., & Min, R. (2024). Intelligent wearable photonic sensing system for remote healthcare monitoring using stretchable elastomer optical fiber. *IEEE Internet of Things Journal, 11*(10), 17317–17329.

8 Collaboration between Universities and City Governments

8.1 University–City Partnerships for Urban Innovation

A collaboration between the city government and universities is connected with an interconnectivity of stakeholder networking: universities, research institutions, government agencies, private enterprise, and community organizations. Alliances such as this share not just ideas but expertise, dynamically setting up the environments where innovation can flower into transformational impacts on urban life. Most of the problems in cities are intertwined, from lack of infrastructure, unsustainability of the environment, proper transportation systems, to finally inequality in housing and socioeconomic classes. These solutions will be interdisciplinary in nature, integrating scientific research into technological innovation with efficient policy frameworks. To this end, research institutions study just the causes of problems inside the urban areas and hence suggest evidence-based solutions. It involves an incrementing of knowledge in think-tank activities that would feed well into policy and new technologies for improved life in cities (Torfing et al., 2024).

The collaboration with industry has thus furnished academia-industry collaboration, if one may say so, with the necessary tools and systems that are required by smart urban management through technological innovations. Capabilities such as smart grids, sensor-based monitoring systems, and data analytics platforms bestowed cities with better resource management, optimization in the use of energy, and real-time observation of environmental conditions (Jonek-Kowalska and Wolniak, 2023; Jonek-Kowalska and Wolniak, 2025). Besides that, some of the state-of-the-art technologies, such as AI, machine learning, and IoT, have also facilitated predictive modelling and decision-making during the designing of resilient and adaptive infrastructure for urban planners. There is also social innovation emerging from collaboration with the local organizations to ensure that technological solutions answer community needs and foster inclusive development (Kempton and Marlow, 2021; Haughton, 2022; Doering, 2021).

This will constitute a supporting structural framework where research and technological development is incorporated into an urban tapestry. By the insight that academic and technological research would provide to policymakers,

DOI: 10.4324/9781003621829-9

regulations and guidelines in encouraging sustainable urban growth should be drafted. They include such tools as creating incentives for green technologies, reinforcing zoning regulations promoting mixed-use development, and implementing transportation policies which can reduce carbon emissions (Pauer et al., 2020). Governance through collaboration models means active involvement in the process; therefore, it is quite transparent and accountable. It highly contributes to maintaining a proper balance between economic growth and equity concerning consideration of the contribution due by the citizens (Zhong and Zong, 2024; Korhonen et al., 2024; Kordel and Wolniak, 2021).

Hotspots of innovation happen at junctures of such effort when these ideas get translated into something tangible, real, and usable. Often, universities also play the role of anchor for these ecosystems by providing an enabling environment with all resources that may be needed for experimentation or developing prototypes. These would include incubators of technology within research enclaves in the city's core that could only help in drawing in the startup enterprises and entrepreneurs to foster the synergy that will nurtures further development. These can be regarded as agglomerations which eventually drive in investments and generate employment opportunities, therefore encouraging the process of entrepreneurship. They also provide an avenue to test new policies and technologies before larger populations are introduced in a city's general setup (Lumbreras et al., 2023; O'Toole et al., 2023; Guenduez et al., 2023).

Such innovation ecosystems technologically update and socially include urban development in doing so. Most of them would focus on the needs of the marginalized communities in filling the gap in access to housing, education, and job opportunities. Scaling up big projects through public–private partnerships mobilizes financial resources and leverages expertise in changing urban landscapes. This all creates sustainable urban systems for long-term growth and resiliency, balancing economic goals with social priorities (Caughman et al., 2020).

University–city partnerships for urban innovation create synergistic frameworks that bring together academics with practical urban applications to help municipal development objectives with research-capable universities. This would place municipalities at the very forefront of responding to the most pressing development challenges requiring data-driven strategies, state-of-the-art technologies, and evidence-based policymaking. This will also broaden the horizon of university research to address practical issues and transform theoretical findings into fruitful outputs with the presentation of updated research methodologies, technological tools, and policy frameworks in serving municipalities by using smart governance, efficient infrastructure management, and sustainable urban growth (Long and Kincaid, 2018; Keeler et al., 2018; Dai et al., 2024). It includes the adoption of smart city technologies, renewable energy systems, and integrated transport systems that will help contribute to attaining sustainable development goals through the improvement of living qualities in cities. In return, this collaboration enables the university to test and further develop theories within application environments, thus building its strength as a producer and disseminator of ideas. Besides, these attract funding for research, facilitation of grants for interdisciplinary work, enhancement of academic reputation, development of experiential learning opportunities, and professional development provided to students and staff. In principle, mutual partnerships further ensure a step-forward stride without fail in order to achieve both municipalities and university goals of development regarding urban building and academic research (Keeler, 2019).

In Table 8.1, there is a description of advantages of university–city partnership in the case of urban innovation.

8.2 Shared Infrastructure and Resource Management

Shared infrastructure and resource utilization between the local government and university campuses are some of the very efficient and effective steps towards sustainability for both. A partnership can be really successful based on supplementing each other's resources and knowledge, providing solutions using innovation that offers cost optimizations, enhances public services, and furthers research and development initiatives. The next important activity of cooperation relates to sharing physical infrastructure. Most universities have several facilities, including research laboratories, libraries, conference centres, and sportive complexes that are not in demand beyond academic hours for most of the day. Many local governments run community centres, parks, and administration buildings that have equally spiky usage. By opening up access to such facilities, both parties avoid parallel investments and are able to extend their respective services to more users, at the same time reducing the environmental impact linked to new constructions. It could also be that a university opens its lecture hall or conference rooms for allowing local governments to hold public meetings or training classes, while municipalities open parks or sports grounds for events organized by a university or student activities (Peng and Xu, 2024; Ficarelli et al., 2024).

Other areas of collaboration will involve the sharing of technological and data resources. Most of the universities are hotbeds of innovation and bound to have advanced IT systems, data analytics equipment, and GIS, which can contribute a lot to the municipalities in terms of carrying out various urban planning, traffic management, and environmental monitoring functions. A joint data-sharing platform can be created wherein the municipalities can gain from academic expertise in enhanced decision-making, disaster preparedness, and sustainable infrastructure. Such that the academic researchers would work with the city planners either to model the impact of climate change or optimize energy use in public buildings so they make informed data-driven decisions on behalf of everybody. It allows universities and local governments to help with the growth of the workforce, especially in lifelong learning. On the other hand, the university could provide courses of training, certification with regard to certain gaps that municipalities will have identified in order to supply a resilient local workforce for challenges and opportunities ahead. On the other hand, municipal governments would grant residents access to

Table 8.1 Advantages of university–city partnerships for urban innovation

Advantage	*Description*
Knowledge exchange	The aim of such partnerships would make a path for the universities to route knowledge into cities and vice versa – municipalities tap into newer research and feed the data and real-world challenges down the line back to universities.
Innovative solutions for urban problems	Some likely applications could be innovations and solutions related to urban issues on traffic congestion, pollution, housing, and sustainability by academic experts.
Enhanced municipal decision-making	The universities provide research-informed inputs to support municipalities in making better decisions with regard to urban planning, policy formulation, and infrastructure management for effective solutions.
Sustainable development	Many of the partnerships are on sustainability that helps cities take environmentally friendly technologies and practices to meet global sustainability goals through a reduction in carbon footprints.
Real-world application of research	It helps universities apply and test theoretical research in the real world of cities. This would facilitate the universities to ensure that their academic knowledge caters to the tangible needs of society and further serves practical development at the city level.
Resource sharing and funding opportunities	Municipalities finance research at the university either financially or with other resources; for that, universities are financed by projects proposed by the cities.
Smart city development	Maybe a smart city technology will eventually be a reason for improvement in living in cities – be it IoT for traffic management or energy-efficient infrastructure.
Community engagement	Most university–city partnerships relate to the community's engagement in designing and carrying out projects in town, making the initiatives responsive to the needs of the people – thus, a reflection of the aspirations of its residents.
Educational opportunities and career development	Students get practical experience through research projects driven by the city in preparing them for careers in aspects like urban planning, public policy, and engineering, among other areas of urban innovation.
Fostering innovation ecosystems	These partnerships represent innovative hubs where academics, governments, industries, and students come together with creative solutions for the urban challenge at hand.
Data-driven urban planning	Universities may also assist a city to collect and analyse data on urban living that helps provide insights for long-term strategic urban planning and city management.
Social impact	At the same time, university–city relationships also bear strong social meanings-from access to affordable housing to disparities in health and public services making a life worthwhile.

Source: Author's own work.

education by funding, scholarships, or transportation. This goes hand in glove with creating not only a strong local economy but also the way to equity in society because of the extended possibilities education and career get (Chen et al., 2024).

Sustainability-wise, this may mean incubating green technologies or sustainable practices within universities. Community partnerships could be developed between a local government and a college or university to pioneer or test renewable energy plans, systems for managing waste, or programs for conservation of water. These projects would also serve practical learning for students while at the same time paying real dividends environmentally and economically for the community. Examples can be university researchers collaborating with municipalities to mount solar panels on public buildings, create urban gardens, or put in place energy-efficient transportation systems (Wang et al., 2020).

Another possible area of collaboration can be through joint research and innovation centres. This in turn creates a path to share resources and talent in solving everything from the local priorities of affordable housing to climate resiliency or public health in mutual innovation centres. These particular areas of hotbed activity usually act as an attraction to other sources of funding, both through government grants and private investment. This also helps in driving longer-term urban sustainability and resiliency through a culture that is creative and fosters experimentation among students, faculty members, and local leaders alike (Pancholi et al., 2020).

The participation in governance and planning ensures coordination between the university and municipality is closer. Advisory committees or task forces comprising members from the two sectors ensure policy and projects address community needs and priorities. Such mechanisms may ensure smooth flow in sharing ideas, identification of emerging challenges, and smoothen the implementation of joint initiatives. This strengthens not only the institutional relationships but also promotes accountability in the use of resources (Kempton and Marlow, 2021).

The sharing of resources by the university and city government can be utilized in their joint efforts towards the implementation of integrated infrastructure wherein the processing of benefits would be facilitated to the academic and local communities. For example, universities and city governments may work in mutual agreement in putting in place mutual transit systems serving students and the rest of the community by buses or even bike-sharing programs. By jointly funding and managing systems, both parties will lessen operational costs while increasing access. This can also take so many private vehicles off the road, thus reducing congestion and, consequently, lowering greenhouse gas emissions. Resource sharing, such as with electric buses or carpool programs, lowers the individual cost and encourages the use of greener, more sustainable modes of transport (Ferraris et al., 2020).

Municipal governments and universities can continue to pursue the use of park, recreational, and community facilities by a joint management agreement for the use of public space. Shared responsibilities could be less costly in keeping the place if both parties have their own ways to increase the use of those facilities. Other fields the universities oriented to academic work may develop in these areas include educative and cultural programs that contribute to the adding of value to the community through its beautification. It can also be approached through green initiatives of community-based collaboration, such as the implementation of an urban garden or other sustainable landscape project beautifying an area by reducing environmental impacts of high levels, for example, the consumption of water and production of waste (Pancholi et al., 2020; Lumbreras et al., 2023).

Utility management is also an important avenue that enables universities and city governments to design and implement better energy, water, and waste management systems. Energy infrastructure can be shared – for example, mutual installations of solar panels or energy-efficient buildings – further reducing the carbon footprint of both with consequential cost savings for both entities. The buying power will be enhanced as well, thereby giving both the ability to negotiate more favourable rates on utilities, hence achieving greater financial efficiencies. These joint waste management programs, in which university and municipal streams are handled together, can meet much larger goals in sustainability and increase recycling rates while reducing landfills. Similarly, joint water conservation projects such as rainwater harvesting systems have helped reduce the use of water and better position both parties to meet environmental regulations (Torfing et al., 2024; Kempton and Marlow, 2021).

The immediate financial benefits of such a collaborative model are evident: shared infrastructure reduces operation costs, joint investments can allow larger-scale projects, and both parties benefit from economies of scale. Ecologically, such collaboration promotes sustainable practices that minimize resource waste, reduce environmental impact, and make the infrastructure system resilient. In return, this eventually leads to some cooperative agreements between universities and city governments to advance the cause of economic and ecological sustainability by building stronger, more connected, and increasingly efficient communities (Haughton, and Han, 2022; Peng and Xu, 2024; Junaid and Ferretti, 2024; Wolniak, 2023).

Shared infrastructure allows sharing resources, hence reducing overall costs to the universities and the local governments by becoming efficient. In addition, through collaboration, there would not be duplication of investments that could have been channelled to transport system, energy management, or public open space, some very capital-intensive functions (Table 8.2). The shared approach would enhance efficiency in the spending of resources whereby public funds are spent with much more effectiveness by benefiting both the academic community and the general public. Concretely, universities and the municipality can work through strategic approaches by having sustainability on campus: developing installations for renewable energy sources, energy-efficient buildings with less environmental impact, that would enhance sustainable urban development (Lumbreras et al., 2023; Ficarelli et al., 2024; Soraggi et al., 2024).

Other benefits of such a partnership are improved services provision. Such shared infrastructures as the public transport systems can also serve both the university staff and the students and the community staying around the university. The collaboration can ensure that those public services are better

Table 8.2 The benefits and difficulties of shared infrastructure

Advantages	*Difficulties*
Shared infrastructure allows for resources to be pooled, leads to a reduction in operational expenses, and avoids investment redundancies, therefore providing considerable cost savings. It sees financial efficiency both within universities and local government.	This is attributed to the fact that universities and the local government have different priorities, thus making coordination very complex in respect of the unifying objectives and timelines.
Sharing infrastructure brings out a much more sustainable operation: energy-efficient buildings and renewable energy solutions that contribute to reduced environmental impact.	Funding responsibility is a source of dispute due to the fact that both parties seek favourable terms, often with vague agreements on financial contributions in the long term.
These, and others, may well provide a means of improving service delivery by universities and city governments to serve public areas and facilities better.	General debates might be at best debatable when issues relating to maintenance and management crop up, especially when the benefit derived from such a transaction bears on either or one of the parties.
Joint infrastructure projects can be innovative, where different approaches and solutions by both sides open up a number of models for advanced infrastructure.	Legal and regulatory barriers, such as many complex issues of zoning variances, liability, and local regulations, may complicate or delay shared infrastructure.
Public spaces, transportation systems, and other community resources are developed through shared infrastructure, in which universities and local governments work together to reinforce community relationships.	That is, shared resources can be overused or underused by one of the parties, having too much reliance on infrastructure that also carries some tension into the relationship.

Source: Author's own work.

maintained and made accessible for more users. Moreover, with collaborative infrastructure projects come innovations. In most instances, universities bring in advanced research and technological know-how, whereas the local governments bring practical experiences that have to do with insight into the needs within a community. This could make a pathway towards innovated advanced infrastructure solutions not easily achievable by separate entities (Wang et al., 2020; Pancholi et al., 2020; Alam et al., 2020).

It can be said that shared infrastructure also involves some challenges. The most important of them are the complexities of coordination by universities and local governments with different goals, priorities, and timelines. It may be rather challenging to make the interests of both parties coincide, especially in a long-term view and long-term decision-making. Financial arrangements may also be one of those contentious issues whereby each party will want to get favourable terms in funding, cost-sharing, and resource allocation. The disagreements over the question of financial contributions and responsibilities complicate the partnership and thus slow down progress (Ruggerio et al., 2021; Soares et al., 2020).

Other than that, shared infrastructure maintenance and management is another challenge. In relation to this is also the question of who pays for upkeep, possibly acting as a source of friction if one party felt that it carries more of the upkeep than the other. Legal and regulatory issues also tend to act as a damper. This will involve tricky navigation by universities and local governments through the different varying laws on zoning, concerns on liability, among others, which may cause the projects to take more time or add to their complexity. Finally, sharing of the resources may further result in unequal use whereby one of the parties utilizes the infrastructure more than what the others do. It could therefore stress the capacity hence strain the partnership (Ficarelli et al., 2024; Chen et al., 2024).

8.3 Social Impact and Community Engagement

The university partnerships, in relation to the development of local governments and businesses in the city, bring about a wide array of social impacts that can be seen through a variety of spheres in society. Probably the most important impact these partnerships create is that they foster an inclusive environment to which education, innovation, and community development relate to one another for the benefit of the large population. Knowledge transfer gives the base for university–city collaboration. The fact that universities are centres for research and learning gives them a privileged position in view of offering values to the local community. The same cities give universities real places where knowledge, as hinted above, can become application and therefore highly relevant for social mobility. The institution developed relationships between universities and cities, access to higher education, mentorship opportunities, and professional opportunities for strategies that reduce inequality on paths to personal and collective betterment. Indeed, the social empowerment aspect is greatly emphatically stressed in those inclusive cities that are supportive of students who are non-traditional or residents (Malhotra et al., 2024, Yao and Chen, 2024).

Higher education partnerships within cultural diversity: Many universities represent centres of creativity and diversity, intellectual pluralism; the presence of a university evokes in an atmosphere of openness, innovation, and critical thinking. The institution may also contribute to the cultural life of the city through the organization of various events-lectures and workshops, art exhibitions, or performances-that draw in city residents and visitors alike interested in its intellectual and artistic activity. Such cooperation between the university and local artists as well as cultural organizations materializes into public arts projects, festivals, and other such events that unite a split populace. Other social implications of cooperation between universities and cities involve civic engagement and community service. Most universities engage their students in volunteer activities and service learning whereby students engage their communities. Besides, they can also be involved in local social issues, local charities, and even government agencies dealing with urban development and policymaking. These give them an added advantage in being responsible citizens post-academics rather than just remaining confined to community service. Better civic engagement builds a strong, better-connected community of citizens themselves, who in this respect are not just passive recipients of services but also contributors to making the environment even better (Timmermans et al., 2023; Birsens, 2024).

This can also lead to further development of social cohesion within the university and in the city. It can tear down barriers between heterogeneous groups of students, staff, residents, and commerce by providing them with platforms of intercultural dialogue and mutual understanding. Thus, universities could eventually make a contribution to the healing of acute social tears in urban life through tolerance and intercultural communication. It could be a society in which differences are celebrated rather than feared, and in which communities start to come together to find common solutions to common problems. Indirectly, there would also be some social consequences of university–city economic impacts. Already, the universities are major employers and contribute massively to the city's economy, but with collaboration with the city, it should rise in job prospects, improving local enterprise growth and investment opportunities in the city. Where large numbers of university students and staff injected into cities, it usually gives vigour to whole neighbourhoods of individual cities, often transforming sections that have been underutilized. This will also result in the betterment of the delivery of public services, infrastructure improvement, and, accordingly, an improved lifestyle for the residents, which will altogether elevate social welfare among the population (Valenzuela Muñoz et al., 2024; Duyen et al., 2023; Wolniak et al., 2024; Wolniak and Jonek-Kowalska, 2022).

This sustainability of university–city partnership might be able to harness a certain degree of social responsibility towards environmental issues. Many universities lead various initiatives in sustainability and related research, and by collaborating with the local governments and organizations, green policy and projects with dual outcomes for environment and community benefit have touched base from encouraging efficiency in energy use to the proper ways of waste disposal and call for sustainable planning of the cities (Brombin, 2023). The result is not only a more sustainable and cleaner city but induction of community ethos that inculcates values into environmental stewardship and long-term ecological health (Table 8.3).

These higher learning institutions can provide a home for centres of knowledge and innovation, as well as a ready avenue for resources: research expertise, educational programs, even a platform for mobilization of dialogue for the resolution of immediate social concerns. Such collaborations create opportunities where students, faculty, and community members work together on projects with community benefits. This would probably lead to increased civic engagement since people are empowered in the solution of problems affecting them as a person (Groulx et al., 2021; Wolniak and Stecuła, 2024; Stecuła and

Table 8.3 The social effects of university–city partnerships

Social effect	*Description*
Knowledge transfer	They give back through research, education, and innovation that provide social mobility and an opportunity for personal and collective improvement.
Cultural enrichment	Events, exhibitions, and performances improve urban cultural life by offering universities as centres of creativity, intellectual exchange, and artistic expression.
Civic engagement	Students and faculty are involved in volunteer activities and service learning in a way that affects the community with service and builds its students into active citizens.
Social cohesion	It promotes intercultural dialogue since universities' collaborative projects and participation by the local communities break the ice in social life and establish a climate of mutual understanding and tolerance, adding to the social environment improvement.
Economic revitalization	This creates university jobs, enhances local business, and attracts investment to assure economic growth for public services that improve the social fabric.
Community empowerment	Such partnerships ensure access to education, mentorship, and professional opportunities that, in turn, reduce inequality and empower locals.
Sustainability and environmental stewardship	The universities have been in the lead in green practices while at the same time supporting local authorities in implementing the green policy as a way of contributing to a healthily sustainable community.
Social integration	This kind of collaboration between university and city provides an environment for interaction and engagement in a diverse array of individuals towards a cohesive society.

Source: Author's own work.

Wolniak, 2022). The university should make it possible for the students to participate in projects that have clear social results; for instance, working in a community-based organization, serving the vulnerable populations, or in the conservation of the environment. As most of these programs provide pragmatic student experience, they deal with the well-being of the people in the community. Such faculty members have been especially interested in community-based research in which persons in the local area were involved in a problem-definitional situation leading up to data collection and remedy. This has the attendant effect of affording confidence that the remedies being realized are rooted in the grassroots of the community, and tend to be relevant and suitable (Masizana and Salubi, 2023; Revington et al., 2023; Wolniak and Skotnicka-Zasadzień, 2021).

University–city partnerships therefore provide a platform for the open discussion of social issues where all interested stakeholders can share their thoughts and experiences with one another and come up with collective goals. University-led public forums, workshops, and seminars bring important social challenges like poverty, education, health care, and inequality to public consciousness.

These, it facilitates dialogues, allowing universities to create an educated and engaged citizenry so that persons are enabled in decision-making processes affecting them. In addition, a university can mediate across gaps among several groupings in the community by pulling together people from across different divides: social, cultural, or economic (Dibbern et al., 2023; Pereira et al., 2023).

Addressing social issues through university–city alliances also involves devising systemic solutions by way of policy research and advocacy. Universities, in their own right, being experts in various fields, can provide much-needed insight into the causes of social problems and recommend evidence-based solutions. Along with local governments and nonprofits, universities can help shape public policy that addresses inequality, social justice, and other critical concerns. This further ascertains that the coming-up approach will guarantee solutions through this collaboration, which bases not only on academic research but also on experiences and needs within the community (Hurenko et al., 2024; Carroll et al., 2023).

Such issues as unemployment, housing, and education are addressed through training, sharing of resources, and supporting people in university–city partnerships in marginalized communities. The universities may thus be in a position to offer programs of job training, mentoring, and internship that provide skills for better-paying jobs. In addition to that, they could link up with schools and organizations within the neighbourhood to help improve educational performance and support students in areas that are less privileged. Long-term changes might be instituted within university–city alliances that were committed to developing local talent and filling gaps in the access of resources (Vallance et al., 2020; Rahmat et al., 2020). Other university–city partnerships can install a culture of mutual concern where problems are confronted as well as remedies are established together to improve life for all citizens: from the leadership of the municipality to academic circles. These alliances encourage active participation and allow for collaboration; in so doing, they address the immediate social issues and lay the groundwork for a more engaged, inclusive, and resilient society. It is through such actions that universities are in positions to act as strong agents of change in the pursuit of communities that are much more equitable, sustainable, and socially cohesive (Ruggerio et al., 2021; McCann et al., 2023; Wolniak, 2024; Wolniak and Skotnicka-Zasadzień, 2018).

For instance, colleges act as civic anchors in metropolitan regions, informing key investments in education, public health, and social innovation. In turn, the role of the college will extend far beyond the classroom as each institution becomes integral in the building of healthier, equitable, and sustainable cities. By that, colleges shape the future through academic, research, and community engagements in the metropolitan areas. It is a setting that is inextricably tied to education, public health, and social development; thus, they help build a living, resistant, and successful urban community (Aguilar et al., 2021; Ou et al., 2024). In Table 8.4, there is a description of main tactics for encouraging significant community engagement in university activities.

Table 8.4 Tactics for encouraging significant community engagement in university activities

Tactic	*Description*
Service-learning programs	It has incorporated community-based projects in their academic programs, enabling students to apply their knowledge in different local social issues and have communities benefit from such initiatives.
Community-partnered research	Faculty and students work with local organizations, governments, and businesses to conduct research about community needs so that any results directly benefit the community.
Public forums and discussions	Organizing events such as lectures, workshops, and panel discussions on issues of relevance to the community stirs up dialogue between the university and the community.
Volunteering and internship programs	Having students volunteer or intern at local nonprofits, government agencies, and businesses creates hands-on experiences for them that promote betterment within the community.
Civic engagement campaigns	Universities lead campaigns that mobilize students, faculty, and staff to participate in activities that range from voting to social activism and to volunteering in the local community.
Collaborative projects	Create a collaborative effort between the organization and the community for a neighbourhood clean-up, educational program, or an arts project that includes the community.
Open campus events	Public events become a mechanism for creating resident engagement with the university through non-threatening, non-exclusive venues: cultural, social and educational.
Community advisory boards	Advisory boards made up of community leaders along with university officials ensure the voices of the community are heard and part of university decision-making processes.
Public–private partnerships	Sometimes, universities partner with local governments and businesses on such urban challenges as infrastructure development, healthcare access and economic growth.
Inclusive programs for diverse populations	These would be further complemented by commitments of the larger community outreach and support programs for immigrants, low-income residents, and people with disabilities.
Cultural and arts outreach	In the sphere of culture and art, events may focus on community-based exhibitions, performances, and music festivals, all of which are especially likely to help involve the residence and to strengthen the university's cultural connection with it.
Local leadership training	These include university-offered leadership training programs that give active roles in local governance, business, and social undertakings to members of the community.

Source: Author's own work.

Community outreach programs instituted are just a couple of ways it has been really effective, in fact, in ways a university might develop this relationship with the community as a force for positive change. Specific examples include: 'The Berkeley Food Institute' was established at the University of California, Berkeley to attend to issues such as community food security, sustainability, and justice. It is through research, advocacy, and hands-on programs that the initiative has been in a position to link students and faculty with underserved populations and food justice organizations. The university collaborates with local schools, food banks, and community farms in a bid to bring access to healthy food to the poor communities while providing experiences for students to solve pressing current social issues. Indeed, this is a great example of linking academic research together with community service to have an impact, sure and real both for learning in students and local well-being (Growing resilience, 2024; Creating, 2024; Fanshel and Iles, 2020; Bartlett, 2023.

Another very successful example of such a program is that at the University of Chicago known as the 'Civic Engagement Program', through which students are matched out to local nonprofit organizations and government agencies to participate in projects that improve urban neighbourhoods' quality of life. Examples of initiatives on which participants can work include urban planning and policy analysis and community development projects dealing with things such as affordable housing, crime prevention, and economic equality. The initiative has allowed students to apply their academic knowledge to real-life problems and, in the process, provided much-needed assistance to organizations working at grass-root levels. This type of project helps the university inculcate a culture of civic responsibility among its students and contributes towards raising a generation of leaders committed to the welfare of their communities (Poulos et al., 2022; Thompson, 2012; Student Voice, 2018).

Another example of community engagement in action is the 'Community Interaction' program at the University of Cape Town: connecting higher education students to communities in South Africa to be of service in some of the major challenges facing the nation-to wit, access to quality educational and healthcare opportunities. This collaboration with schools, clinics, and other NGOs provides students with a facility for the conversion of mere theoretical knowledge into practice in providing direct service within communities. Probably the greatest achievement which this model is capable of gaining is how it is capable of connecting academic learning with applied practice that does give real results to those most in need along with a heightened academic value (Millan et al., 2018' Il-haam et al., 2022; Lavack, 2021).

References

Aguilar, S., Kean, E., Kushner, D., Larsen, P., & Schlegelmilch, J. (2021). Partnerships to enable smart cities and support resiliency. In *2021 IEEE conference on technologies for sustainability, SusTech 2021* (p. 9467451).

Alam, G.M., Al-Amin, A.Q., Forhad, A.R., & Mubarak, M.S. (2020). Does the private university sector exploit sustainable residential life in the name of supporting the fourth industrial revolution? *Technological Forecasting and Social Change, 159*, 120200.

Bartlett, B. (2023). *Berkeley Food Utility and Access Resilience Measure (FARM)*. https://berkeleyca.gov/sites/default/files/legislative-body-meeting-attachments/2023-05-08%20ITEM%2002%20SUPPLEMENTAL%20%28PRESENTATION%29.pdf, (Accessed 15.01.2024).

Birsens, J. (2024). Analysing urban integration through place attachment: How do university students contribute to the formation of an integrated urban space?, *Population, Space and Place*, *30*(7), e2783.

Brombin, A. (2023). Bubbles of social contact in interstitial spatialities. Interactional dynamics among university students in the city of Padua, Italy. *Etnografia e Ricerca Qualitativa*, *16*(2), 343–352.

Carroll, C., Fitzgibbon, I., & Caulfield, M. (2023). Community and university partnerships: Integrating sustainable development goals 3, 4, 10, 11, and 17, *International Journal of Speech-Language Pathology*, *25*(1), 102–106.

Caughman, L., Keeler, L.W., & Beaudoin, F. (2020). Real-time evaluation of city–university partnerships for sustainability and resilience. *Sustainability*, *12*(21), 1–18, 8796.

Chen, Q., Xia, X., Hui, Y., & Zhang, S. (2024). Urban intellectual property strategy and university innovation: A quasi-natural experiment based on the intellectual property model city of China. *Systems*, *12*(1), 21.

Creating the future of sustainable and just food systems (2024). Berkley Food Institute, https://food.berkeley.edu/wp-content/uploads/2020/05/2020-BFI-Fundraising-Overview.pdf (Accessed 15.01.2024).

Dai, L., He, Y., & Luo, S. (2024). Does the innovative city pilot policy improve university–industry collaborations in China?, *Journal of Technology Transfer*, *50*(1), 375–398.

Dibbern, T. A., Bertuluci, F. B., Cristofoletti, E. C., Serafim, M. P., & Da Costa Ferreira, L. (2023). The university contribution to sustainable development in Brazil: An analysis of the institutionalization of HIDS-Unicamp. *Desenvolvimento e Meio Ambiente*, *62*, 372–388.

Doering, C., Schmidtner, M., & Timinger, H. (2021). Collaboration for innovation between universities and smart cities. In *2021 IEEE European technology and engineering management summit, E-TEMS 2021 - conference proceedings* (pp. 82–86).

Duyen, N. T. T., Nho, T. T. N., Loan, P. T., Toan, D. X., & Hang, H. T. X. (2023). Factors affecting adaptability of teachers in teaching at the university of social sciences and humanities, vietnam national university ho chi minh city under the Covid–19 pandemic. *Journal of Educational and Social Research*, *13*(5), 236–254.

Fanshel, R. Z., & Iles, A. (2020). Transforming the campus foodscape through participatory mapping. *Case Studies in the Environment*, *4*(1), 1120325.

Ferraris, A., Belyaeva, Z., & Bresciani, S. (2020). The role of universities in the Smart City innovation: Multistakeholder integration and engagement perspectives, *Journal of Business Research*, *119*, 163–171.

Ficarelli, L., Vacca, V., & Turchiarulo, M. (2024). University, city and territory. Caserma Magrone as a new Innovation Park. *TECHNE, (28*), 211–220.

Groulx, M., Nowak, N., Levy, K., & Booth, A. (2021). Community needs and interests in university–community partnerships for sustainable development, *International Journal of Sustainability in Higher Education*, *22*(1), 274–290.

Growing resilience, 2021–2024 impact report, (2024). Berkley Food Institute. https://food.berkeley.edu/wp-content/uploads/2024/05/Berkeley-Food-Institute_Growing-Resilience-2021%E2%80%932024-Impact-Report_MAY24_8.5x11_DIGITAL_FINAL_spreads.pdf (Accessed 15.01.2024).

Guenduez, A. A., Frischknecht, R., Frowein, S. C. J., & Schedler, K. (2023). Government-university collaboration on smart city and smart government projects: What are the success factors? *Cities*, *144*, 104648.

Haughton, N. A., & Han, D. (2022). Internationalisation through a Confucius sister city partnership: Examining a 10-year sister city and university bi-lateral partnership. *Compare*, *52*(8), 1277–1295.

Hurenko, O., Suchikova, Y., Kravchenko, N., Nesterenko, M., & Petryk, K. (2024). Employment of young people with disabilities: The potential of social partnership of universities, municipalities and the labor market of Ukraine. *Work (Reading, Mass.)*, *79*(3), 1407–1423.

Petersen, I. H., Kruss, G., & van Rheede, N. (2022). Strengthening the university third mission through building community capabilities alongside university capabilities. *Science and Public Policy*, *49*(6), 890–904,

Jonek-Kowalska, I., & Wolniak, R. (2023). *Towards sustainability and a better quality of life?* London: Routledge:.

Jonek-Kowalska, I., & Wolniak, R. (2025). *Exclusion in smart cities assessment and strategy for strengthening resilience*. London: Routledge.

Junaid, M., & Ferretti, M. (2024). Planning and designing a sustainable mobility system in rural areas. In *Advances in science, technology and innovation* (pp. 205–228).

Keeler, L. W., Beaudoin, F., Wiek, A., Kay, B., & Forrest, N. (2019). Building actor-centric transformative capacity through city-university partnerships, *Ambio*, *48*(5), 529–538.

Keeler, L. W., Beaudoin, F. D., Lerner, A. M., Wiek, A., & Lang, D. J. (2018). Transferring sustainability solutions across contexts through city-university partnerships. *Sustainability*, *10*(9), 2966.

Kempton, L., & Marlow, D. (2021). Realising the potential of universities for inclusive, innovation-led development: The case of the Newcastle City Futures Urban Living Partnership pilot. In *Hope under neoliberal austerity: Responses from civil society and civic universities* (pp. 169–185).

Kordel, P., & Wolniak, R. (2021). Technology entrepreneurship and the performance of enterprises in the conditions of Covid-19 pandemic: The fuzzy set analysis of waste to energy enterprises in Poland. *Energies*, *14*(13), 3891.

Korhonen, T., Salo, L., Reinius, H., Lavonen, J., & Hakkarainen, K. (2024). Creating transformative research–practice partnership in collaboration with school, city, and university actors. *Education Sciences*, *14*(4), 399.

Lavack, V. (2021). *Scholarship of engagement practices are so divergent across universities that this necessitates a dedicated focus*, https://usaf.ac.za/scholarship-of-engagement-practices-are-so-divergent-across-universities-that-this-necessitates-a-dedicated-focus-vivienne-lawack/ (Accessed 15.01.2024).

Long, J., & Kincaid, P. (2018). A red city goes green: The renewable energy partnership of Georgetown, Texas and Southwestern University. *Sustainability (United States)*, *11*(6), 298–306.

Lumbreras, J., Moreno-Serna, J., Palau, G., Sánchez-Chaparro, T., & Mataix, C. (2023). University–city partnerships for sustainable urban transformations. In *The Bloomsbury handbook of sustainability in higher education: An agenda for transformational change* (pp. 279–295).

Malhotra, S., Anil, K., & Kaur, A. (2024). Impact of entrepreneurial ecosystem on sustainable social enterprises - A mixed-methods research. *International Journal of Experimental Research and Review*, *41*, 206–216.

Masizana, F., & Salubi, O. (2023). (Un)Social media services at the people's university: A survey of City of Cape Town Public Libraries, *South Africa, Internet Reference Services Quarterly*, 27(4), 247–267.

McCann, L., Hutchison, N., & Adair, A. (2023). The role of UK universities as economic drivers in a localisation agenda: A case study of City Deals. *Land Use Policy*, *134*, 106938.

Millan, J. M., Goodman, S., & Schmid, B. (2018). *Illuminating "Transaction Spaces" in higher education: University–community partnerships and brokering as "Boundary Work"*. https://uct.ac.za/sites/default/files/content_migration/uct.ac.za/1681/files/JM%2520SG%2520%2526%2520BS_Transaction%2520spaces_final%2520%25281%2529.pdf (Accessed 15.01.2024).

O'Toole, T. P., Holl, J., & Freyder, P. (2023). A community partnership in service to the homeless: University of Pittsburgh and the City of Pittsburgh. In *Creating community-responsive physicians: Concepts and models for service-learningin medical education* (pp. 115–119).

Ou, Y., Zhen, Z., & Cai, R. (2024). Structural efficiency evaluation of humanities and social sciences research based on MEA model: A case of universities in Jiangsu, China. *SN Social Sciences*, *4*(4), 83.

Pancholi, S., Yigitcanlar, T., Guaralda, M., Caldwell, G.A., & Medland, R. (2020). University and innovation district symbiosis in the context of placemaking: Insights from Australian cities, *Land Use Policy*, *99*, 105109.

Pauer, S.U., Pilon, A., & Badelt, B. (2020). Strengthening city–university partnerships to advance sustainability solutions: A study of research collaborations between the University of British Columbia and City of Vancouver, *International Journal of Sustainability in Higher Education*, *21*(6), 1189–1208.

Peng, C., & Xu, H. (2024). Universities and cities: The impact of higher education on urban innovation, *Entrepreneurship, and Economic Growth, Asian Economic Papers*, *23*(2), 33–56.

Pereira, D., Leitão, J., Oliveira, T., & Peirone, D. (2023). Proposing a holistic research framework for university strategic alliances in sustainable entrepreneurship, *Heliyon*, *9*(5), e16087.

Poulos, Ch., Arenas, I., & Lewis, A. (2022). *Changing the frame: Civic engagement through a racial equity lens, University of Illinois at Chicago*. https://www.cct.org/wp-content/uploads/2022/08/IRRPP_300dpi_Civic-Engagement-Through-A-Racial-Equity-Lens.pdf, (Accessed 15.01.2024).

Rahmat, H., Marshall, N., Steinmetz, C., Thompson, S., & Corkery, L. (2020). The role of smart city initiatives in driving partnerships: A case study of the smart social spaces project, Sydney Australia. *Contemporary Urban Design Thinking, Part F2*, 143–159.

Revington, N., Zwick, A., Hartt, M., & Schlosser, J. (2023). Universities and urban social structure: Gentrification, studentification, and youthification in five United States legacy cities. *Urban Geography*, *44*(1), 83–104.

Ruggerio, C. A., Querejeta, G. A., Conicelli, K. B., & Lombardo, R. J. (2021). Integration of municipal state, society and university efforts for sanitary risk prevention associated with Aedes aegypti mosquito in the metropolitan area of Buenos Aires, *Argentina. Tropical Medicine and International Health*, *26*(7), 789–799.

Soares, I., Weitkamp, G., & Yamu, C. (2020). Public spaces as knowledgescapes: Understanding the relationship between the built environment and creative encounters at dutch university campuses and science parks. *International Journal of Environmental Research and Public Health*, *17*(20), 1–30, 7421.

Soraggi, D., &Campanini, F. (2024). Applying 4.0 technologies to public spaces. exploring new functions and interactions in Savona University Campus. In *Lecture notes in civil engineering, 467 LNCE* (pp. 157–168).

Stecuła, K., & Wolniak, R. (2022). Advantages and disadvantages of e-learning innovations during COVID-19 pandemic in higher education in Poland. *Journal of Open Innovation: Technology, Market, and Complexity*, *8*(3), 159.

Student Voice, Engagement & Civic Life Component (2018). *CPS school excellence framework*.https://centerx.gseis.ucla.edu/wp-content/uploads/2019/01/Chicago-Public-Schools-SEF-Civic-Life-Guidebook-2018.pdf (Accessed 15.01.2024).

Thompson, J. (2012). *Dialogue and deliberation for civic engagement in Chicago: Building a community of practice, university of Illinois at Chicago*. https://ipce.red.uic.edu/wp-content/uploads/sites/41/2018/10/DDSURVEYReportFinal1_15_13.pdf (Accessed 15.01.2024).

Timmermans, R., Naccarato, K., Salvato, M., Stidham, J., & Wen, Y. (2023). Space-based solutions for smart cities. In *Proceedings of the international astronautical congress, IAC*, 2023 October.

Torfing, J., Sivertsen, H., & Torvatn, H. (2024). Metagovernance of co-creation in city–university partnerships: How to avoid being stuck in the middle?, *Public Administration*, *102*(4), 1711–1727.

Valenzuela Muñoz, A., Ruiz Huaraz, C.B., Valenzuela Ramos, M. R., & Lioo Jordan de Baldeos, F.D.M. (2024). Satisfaction of a social program received by students of a university in the city of Huacho. *Atencion Primaria*, *56*(9), 103011.

Vallance, P., Tewdwr-Jones, M., & Kempton, L. (2020). Building collaborative platforms for urban innovation: Newcastle City Futures as a quadruple helix intermediary, *European Urban and Regional Studies*, *27*(4), 325–341.

Wang, X., Ren, Y., Zhang, L., & Ma, Y. (2020). The influence of technological innovation ability of universities on regional economic growth: Take coastal provinces and cities for example. *Journal of Coastal Research*, *103*(sp1), 112–116.

Wolniak, R. (2024). Hospital distribution in Polish provinces as a factor of smart living. *Economics & Sociology, Centre of Sociological Research - Ukraina*, *17*(1), 132–150.

Wolniak, R., & Skotnicka-Zasadzień, B. (2018). Developing a model of factors influencing the quality of service for disabled customers in the condition s of sustainable development, illustrated by an example of the Silesian Voivodeship public administration. *Sustainability*, *10*(7), 2171, 1–17.

Wolniak, R., &Skotnicka-Zasadzień, B. (2021). Improvement of services for people with disabilities by public administration in Silesian Province Poland. *Sustainability*, *13*(2), 967, 1–16.

Wolniak, R. (2023). European Union smart mobility—aspects connected with bike road system's extension and dis-semination. *Smart Cities*, *6*, 1009–1042.

Wolniak, R., & Stecuła, K. (2024). Evaluation of quality of innovative e-learning in higher education: An insight from Poland. *Applied System Innovation*, *7*(6), 109. https://doi.org/10.3390/asi7060109

Wolniak, R., Gajdzik, B., Grebski, M., Danel, R., & Grebski, W.W. (2024). Business models used in smart cities—theoretical approach with examples of smart cities. *Smart Cities 7*, 1626–1669.

Wolniak, R., & Jonek-Kowalska, I. (2022). The creative services sector in Polish cities. *Journal of Open Innovation: Technology, Market, and Complexity 8*, 17.

Yao, Y., & Chen, Z. (2024). Shuttle bus scheduling optimization based on spatio-temporal networks. In *Lecture notes in electrical engineering, 1209 LNEE* (pp. 259–271).

Zhong, Z., & Zong, J. (2024). Universities as cities of flows: Decoding cross-regional university partnerships for sustainable development in China, *International Journal of Comparative Education and Development*, *26*(3), 208–225.

9 Challenges and Opportunities in Creating Smart Campuses

9.1 Financial and Technological Hurdles

Implementation of the smart university campus is to be preceded by the prudent consideration of a set of technical and financial issues. Among the major challenges, high costs are involved in building new infrastructure or scaling up the existing infrastructure to cater for the technological needs arising in the environment of a smart campus. The depth of this development may call for investment in deep and very deep hardware and software in designing and building such new modernized spaces that can absorb this advancement. Preliminary investment cycles for integrating those structures with IoT technologies, higher bandwidth connectivity, newly equipped data centres, and power over economy may involve very high costs (Abulibdeh, 2024).

These financial issues are circling not only around the initial capital expenditure but also around continuous operation and maintenance expenses. All the sophisticated technologies deployed in a smart campus need continued monitoring and updating; for all this, further technical support needs to be provided. Cybersecurity measures that guarantee data privacy and patch vulnerabilities in software require further qualified manpower and resources, thus enhancing the financial burden. Additionally, since the technologies are ever-changing, institutions will have to budget periodically for upgrade and replacement to avoid the same from getting outdated and unusable in the long run (Bellaj et, al., 2024).

The issues of finance aside, there are a number of technical problems in integrating the smart systems. Most universities have been built upon older systems that were not designed to operate in a modern environment. The compatibility of the older infrastructure with the new digital tools creates some tricky interoperability issues that demand customized solutions and longer implementation. There are a lot of systems operating on one platform – usually quite hard to manage technically without causing disruption, including but not limited to access control, energy management, smart lighting, and digital learning platforms (Xu and Zhao, 2024; Dominguez-Botano, et al., 2024).

These latter are exacerbated by the technical challenges associated with managing those volumes of data produced in those connected systems themselves for effective and efficient analytics, using that data in decision-making,

DOI: 10.4324/9781003621829-10

performance optimization, and predictive maintenance; similarly highly sophisticated analytical tools and capabilities exist in artificial intelligence. Deploying such systems requires expertise and resources not necessarily available to each and every institution, especially with extremely tight budgets. Other than that, universities must take into consideration human factors stifling the implementation of smart campus solutions. The necessary training of faculty, staff, and students for the use of new systems is badly needed. Yet, resistance to change restricts this. Designs must be user-friendly and interfaces intuitive in order to reduce the learning curve and encourage use (Bandoria et al., 2024). Several of the major critical technical and financial challenges related to the creation of smart campuses are compared in Table 9.1.

Most universities face serious financial difficulties to make their campuses smart. Possible financing sources will include the following. Among the funding sources, which are pretty common and feasible, one would include different government grants and subsidies since many public authorities are well aware of this need for modernizing educational infrastructures to be at par with the rapidly changing demands caused by technologies. This has made many ideas on sustainability, digital transformation, and research-driven projects, among others, apt for smart campus development funding. It further allows universities to take advantage of national and regional innovation programs in energy efficiency and digital inclusion, reducing the burden of high upfront costs (Chen et al., 2024a; Famoriji and Shongwe, 2024; Eldeeb and Alves, 2024).

Another promising avenue of financing comes through corporate collaboration and industry partnerships. Most of the technology companies are very much interested in investing in educational institutions to demonstrate their products, run pilot programs, and establish long-term relations with the future professionals. These private sector partners, in turn provide funding, technical expertise, and access to modern technologies in return for these strategic alliances. These are sponsorship forms, joint ventures, and co-funded research arrangements in which the sharing in the development costs, with attendant corporate knowledge and resources benefits, may be shared by the institutions. This provides them the unique benefit of often trying and perfecting their technologies in real-time surroundings; hence, it generally is a mutual benefit model (Roda-Sanchez et al., 2024' Alvarez et al., 2024).

Other than grants, the investment model for partnerships, universities will consider where there are long-term revenue generation models or cost-sharing mechanisms. Public–private partnerships will provide for creating mechanisms through which institutions can avail themselves of private investors for infrastructure improvement loans defrayed on a long-term basis through operational revenues or leasing agreements. For example, universities can use the Performance-Based Financing model in which funding is tagged to certain metrics like energy savings or other operationally efficient metrics. This would either reduce any huge upfront capital need by a great margin and lever long-term benefits against initial costs (dos Santos, et al., 2024).

Table 9.1 Technical and financial difficulties in smart campuses creation

Technical difficulty	*Description*	*Financial difficulty*	*Description*
Infrastructure modernization	Upgrade or replace the old legacy systems to accommodate the support of smart technologies and assure compatibility.	**High initial investment**	The capital investment in new infrastructure, hardware, and software integration alone.
System integration	Ensure that communications across diverse systems – from IoT devices through energy management and security platforms – operate seamlessly.	**Maintenance and operational costs**	Ongoing expenses for update packages to the software, cybersecurity, technical help, and troubleshooting.
Data management and analytics	Use systems that process and analyse large volumes of data transmitted from connected devices.	**Software and platform licenses**	Licensing of different advanced analytical tools and platforms.
Cybersecurity implementation	Create strong security protocols that reduce the probability of data breaches or cyber threats.	**Cybersecurity investment**	Security systems, data protection, and compliance with relevant privacy legislation.
Scalability and future proofing	Design systems that can accommodate change due to technological evolution and increased demand over time.	**Technology upgrades and replacements**	Recurring costs to refresh obsolete technologies and replace old hardware and software.
User training and adoption	Training of faculty, staff, and students on efficient use of smart campus systems; how to handle resistance to change.	**Training and development programs**	Allocation of costs within training programs, workshops, and instructional materials to end users.
Energy efficiency and sustainability goals	Implementation of sustainable energy solutions, optimization of energy usage without compromising performance.	**Energy management systems installation**	Invest in smart energy systems, sensors, and renewable energy technologies.

(*Continued*)

Table 9.1 (Continued)

Technical difficulty	*Description*	*Financial difficulty*	*Description*
System reliability and redundancy	Reliability within operation of systems with backup mechanisms for non-disruption or failure.	**Backup infrastructure investment**	Cost of redundant systems, backup servers, and disaster recovery plans
Vendor dependence and customization	Balancing third-party solutions with custom development to meet very specific needs yet not tie into vendor lock-in.	**Vendor contracts and service agreements**	Long-term contracts with third-party vendors on maintenance, support, and licenses.

Source: Author's own work.

Others include bond issuances or availing of low-interest loans tailor-made for educational institutions. Most governments and financial institutions give very favourable lending terms for sustainability and smart technology-related projects, which allows universities to spread their costs over extended tenures. This would lessen the immediate financial burdens while the institutions are still capable of investing in large-scale upgrades and innovations. Other than this, large-scale endowments and alumni donations for modernization may also be tapped if the campaigns clearly underscore the message that the impact of smart campuses would be long-range in their improvement of the aspects relating to educational quality, sustainability, and students' experiences (Li and Chen, 2024; Pexyean et al., 2024).

Universities can also subscribe to technology, rent equipment and infrastructure, while making the least outright purchases. This will not only reduce initial costs but also leave room for upgrading and scaling up while technologies evolve. Most of the financial barriers to smart campus development can be overcome by applying these financing strategies in concert, thus maintaining competitiveness and the ability to deliver world-class education in this increasingly digital age (Delnevo et al., 2024).

That will surmount many of the important technical hurdles in the path to make the campus smart-strategic and multi-pronged modernization of the infrastructure, integration of the systems, data management, and cybersecurity. For instance, state-of-the-art technologies that are upgraded over the older legacy systems would be achievable upfront. That could be enabled by modular and scalable solutions that provide for gradual integrations, which will allow integrating existing operations in a non-destructive way. It is the Universities that must invest in such open standards and interoperable platforms to reduce the complexities of interlinking of various kinds of systems with IoT devices, energy management tools, and digital learning environments (Stecuła and Wolniak, 2022; Wolniak and Stecuła, 2024). Such management would be easier by using a unified management suite that would give single consolidated interface to operate and analyse a host of campus-wide technologies (Florentin et al., 2024).

That said, for such, managing data has grown so complex that investment in multimodal analytics platforms with expanded cloud storage for scalability and flexibility is a must. The capability for processing and analytics can further be enhanced by the use of AI and machine learning tools, which would enable the institutions to carry out predictive maintenance, real-time monitoring, and data-driven decision-making. Besides that, data governance frameworks and policies developed further ensure their adherence to the set regulations with regard to privacy, hence protecting sensitive information while offering complete transparency and accountability (Hussien et al., 2024; Baba et al. 2024).

Of equal importance is the address on cybersecurity. This should relate to security policy development, encryption of information, multi-factor authentication, and continuous network monitoring to protect sensitive information. Other ways to enhance protection involve periodic cybersecurity audits and training for faculty and students. Development of incident response plans in order to quickly respond uniformly on the containment of a security breach and ensure minimal campus disruptions is also an add-on (Venkatesan et al., 2024).

The University shall adopt only such architecture that can accommodate the evolving technologies and requirements, future proofing the systems. Along with scaling efficient infrastructure due to increasing deployment on campuses, the dependency on hardware is reduced by Software-defined networking and Virtualization. Technical support, best practices, and leading-edge innovation of technology providers and experts will lead them through these changing times (Hassanain et al., 2024).

Training and Change Management: The technical impediment becomes less of an issue in comparison to how well the solution will be adopted by its users. Facets that universities can invest in include workshops, tutorials, and online resources to familiarize faculties, staff, and students with new technologies. A well-thought-of design of the system makes the system intuitive and user-friendly, reduces friction, and makes it thus easier to transition into, hence assuring higher engagement (Kim and Lee, 2024).

Besides, collaboration with technology companies, consultancies, and research institutions will help to develop and install Smart Campus technologies. Similarly, the adoption of the university in the pilot and phase programs involves using these solutions in small scales where they can detect the problem points to make the required amendments before final implantation. These approaches greatly reduced risk and ensured technologies meet their actual needs (Mariappan et al., 2024).

Most universities worldwide have survived a lot in terms of the development of smart campuses, as far as technical and financial issues are concerned, hence coming up with various models from which others could seek inspiration. A great example is that of the University of Eindhoven in the Netherlands; it has an overall strategy for innovation, sustainability, and seamless learning on smart campus. Major challenges are of a technical nature and involve integration into the infrastructure on campus, especially for older buildings that never planned on adapting to these kinds of smart systems. The university tried taking advantage of this by installing IoT sensors installed while upgrading to smarter, more robust on-campus Wi-Fi backbone technologies. University Eindhoven was given the financially relevant governmental grant to fund the green project or energy-saving technologies. In the formed partnership, Philips developed smart lighting and investment systems together with sharing technological companies on the campus of the university. Any complication technologically was resolved through steps of implementation as this modernistic university continued its transition into an adapted and friendly facility to all upcoming new technologies or research needs (Valks et al., 2020; Molenaar, 2020; Eindhoven, 2024).

Another great example of overcoming not only technical but also financial challenges in developing smart campus environments is Nanyang Technological University, Singapore. For example, while creating the Smart Campus of NTU, it was required to engage in a complete rethink of the entire design and its integration with intelligent mobility solutions, real-time campus monitoring, and advanced facilities for research. On the other hand, again, the challenge was facing how NTU would make their systems scalable and interoperable at the rate with which technology evolved. As it has, NTU worked with no more than four–five serious providers of technology and research in-house; developed customized solutions; needing a solution that would give total autonomy for shuttle buses; and also classrooms which should be smartly equipped with AI to a great degree. Under these circumstances, NTU was also economically able to find a route from government initiatives, inclusive of its Smart Nation initiative, to strategic partnerships with the corporations of the world in investments in technology and sustainability for the long term, public fund mobilization, corporate investments, and the probable opportunity to overcome the financial complexities inherent in this kind of projects (Smart Campus, 2025; Smart Campus, 2023; Das et al., 2022).

The second case study is the University of Melbourne, Australia, having overcome certain technical and financial issues. The university wanted its campus to become smart, connected, and help the university fulfil its demand for digital learning and research. Firstly, there was the enormous technical challenge of integrating this large geographically dispersed campus into a single digital infrastructure. This the university addressed through investments in cloud computing and data analytics platforms to manage and integrate its data centrally across departments. At the level of finance, the University of Melbourne derived funding from a combination of state and federal government allocation, private donation, and investment. It was decisive that performance-based financing and savings obtained by the university due to energy efficiency measures were reinvested in further upgrades, thereby softening the financial burden decisively. In these respects, the University will be in a position to enhance the smart campus infrastructure at the front with the least financial burden on its end (Sustainability, 2024; Polin et al., 2024; Chagnon-Lessard et al., 2021; Wolniak, 2023).

Another good example of how a university could overcome several technical and financial difficulties standing in its way to becoming an intelligent campus is the University of Hong Kong, HKU. Some of the major objectives of setting HKU included making its campus more sustainable with several smart integrations like energy efficiency, smart classrooms, and also an advanced campus management platform. The university would be able to surmount such wide-ranging technical challenges only by adopting one central management system that pulls everything – lighting, heating, and security – onto a single platform. Major funding for HKU came from government-backed initiatives and solid industry partnerships, while co-developing and co-implementing smart solutions were made with active engagement of technology providers like Huawei and Google. Similarly, government grants encouraged the work of higher education institutions toward sustainable projects. After careful planning and with partnership, HKU overcame formidable technical and financial challenges to finally become a model 'smart campus', rich in educational experience and environment-friendly sustainability (The Road, 2024; Smart Campus Development, 2024; Report of Quality, 2024).

9.2 Cultural and Social Resistance

Probably, smart campus initiatives would be the most important leap toward modernization in terms of educational settings, operational efficiency, and developing student experiences for colleges. However, at the same time, cultural and social resistance to such advanced technologies as IoT devices, artificial intelligence, and data analytics derived from them is strong enough to forbid a number of institutions from effectively embracing this kind of initiative. Such resistance is partly rooted in phobias: privacy, security of data, enslavement by technology, and even the erosion of traditional academic values (Liu, 2024a; Geng et al., 2024).

Thus, in the present domain, the two sources of cultural opposition are hitched to misgivings related to privacy and surveillance concerns. Most of smart campus technologies is built out of an extended set of processes that involve capturing data, its monitoring using a smart camera, sensor access control amongst others. Whereas that set of changes is targeted for the most part at those facets of safety and resource management relating to students, faculties, and staff, others view these very changes as intrusive because the overriding sense was that they have been closely monitored and thus had lost their anonymity in the process. This, up until now, generally elicits discomfort, which, in turn, emerges into resistance toward setting conditions around collecting and storing data or using these resources. They may even start to wonder if the facility is capable of ensuring that data will not be breached and misused, thus being more incredulous (Kurniawan et al., 2025; Liang and Gong, 2025).

Social opposition can be another explanation of such a generational gap in attitudes towards technology. While to the young student generation, new and innovative things might seem intuitively natural, for seniors both from the faculty and administrative staff, all these novelties are so often hard to digest and grasp simply because of the rapid pace at which technology is evolving around them. For them, the move to smart campus initiatives would mean an interruption of traditional ways and workflows that had existed for a long time, accompanied by anxiety over job security, skill obsolescence, and erosion of human-centred interactions. All this is compounded by the fear that such over-reliance on technology would depersonalize education, with automated processes and algorithms taking over from meaningful human contact and interaction (Wang et al., 2024; Bitencourt et al., 2024).

Sometimes this begets, or at least reinforces, a more philosophical critique rooted in the concern that these kinds of smart campus initiatives tend to further commercialize education. Going even further, many of these critics would argue that the said technologies promote efficiency, profitability, and data-driven decisions rather than intellectual freedom and creativity. Such actions can potentially bring accusations against colleges, considered strongholds of critical thought and cultural development, and put corporate interests ahead of academic integrity. This may alienate those faculty and students who do not consider education as a utilitarian enterprise but as a holistic, humanistic enterprise (Yin et al., 2024).

Resistance may also be supported on grounds of socio-economic inequalities amongst the students themselves, since the implementation of smart technologies runs the risk of increasing such inequalities. For example, students from low-income backgrounds may feel that the pre-condition of access to digital resources for the purpose of participating in academic and co-curriculum activities marginalizes them. High-tech systems will further pin the digital gap: a campus environment in which the digitally savvy are privileged, with those having poor access to devices or with a lack of needed digital literacy skills being at a disadvantage. The result of this will be that people become resentful and embitter the opposition to the adoption of smart campus initiatives (Yu and Zhang, 2024; Farhan and Daghal, 2024).

Other implementation barriers may be religious or cultural in value when some groups consider the technologies of advancement with particular suspicion, perceiving them as factors in direct opposition to a belief or value considered incompatible with traditional ways of learning and interacting (Table 9.2). Indeed, the highest degree of resistance may be felt within those communities that place the highest cultural values on the importance of face-to-face interaction and personal relationships in thought-in other words, the impersonal nature of these digital tools undermines the cultures (Table 9.2; Polin et al., 2024; Lin et al., 2024).

What most are actually scared about is that it will simply change the dynamics of higher education, hence laying claim to job security, privacy, and cultural identity for these institutions. The major fears revolve around job displacement. Essentially, with the smart campus embracing automation, artificial intelligence,

Table 9.2 The cultural and social problems that exist while putting smart campus initiatives into practice

Problem	*Description*
Privacy and surveillance concerns	In fact, most intelligent technologies involve the collection and monitoring of data, which instils fears of personal privacy invasion and constant surveillance. Indeed, students and staff may ask themselves how their data is stored, used, and protected against data breaches.
Generational resistance	Older faculties and staff may object to the new technology based on difficulties in adapting to digital tools, job security, obsolescence of skills, and the loss of traditional methodologies in teaching.
Depersonalization of education	Furthermore, critics argue, it would shrink human interaction and further damage a good, well-meaning relationship between the student and teacher-the atmosphere is not going to be so friendly.
Commercialization of education	Smart campus initiatives would also be liable to focus on profit and efficiency and corporations' interests rather than on academic freedom, intellectual exploration, and cultural values preservation within education.
Socio-economic inequalities	Advanced technologies can also be a source of increasing the digital divide: students from low-income backgrounds could not have access to a device or digital literacy, and hence become unequal in participation and learning opportunities.
Technological dependency fears	Risk of overreliance on technology: these systems can fail, data will be lost, and cybersecurity creates vulnerabilities that will put academic and administrative functions into precarious situations.
Cultural and religious opposition	This might suggest that some cultures and religions even feel that dependence on intelligent technologies is in conflict with their values, which promote face-to-face interaction and conventional means of communication not being impersonal and devoid of feeling.
Lack of trust in institutions	More to the point, however, is that there are stakeholders who will question the ability of colleges to manage this in an ethical manner and their safeguarding of private information-thus the call for transparently articulated policies and assurances above data governance and data transparency.
Impact on academic integrity	There is a fear that analytics-based decisions would supplant both critical thinking and creativity, smart technologies emphasize efficiency over intellectual freedom and depth in education.
Resistance to change	Generalized resistance may relate to the fear of something unknown, the discomfort of modifying well-established routines, and the fear that new systems may create glitches or inefficiencies during the process of assimilation.

Source: Author's own work.

and insights-driven systems to run operations ranging from administrative, security, and teaching methodologies, it may seem as though their roles are in jeopardy to both the faculty and staff. For instance, administrative staff fear responsibilities because some schedules, bookkeeping, facility management that required many hands are being outsourced to automation systems and therefore apparently have nothing to do. But what this points to is also apprehension that could develop in the tutors, taking the virtual classrooms or AI virtual tutoring as a means of easing the need for human resources into the traditional paradigm of teaching, hence reducing contact with humans altogether. In most instances, such beliefs might get even worse, perhaps institutions may never invest in retraining or upscaling their staff to operate with dynamically changing technologies. One can only fathom the amount of data that comes into play when a smart campus works. Major technologies that can be included are variants of surveillance cameras, biometric sensors, and even sensors used to monitor movements within the campus boundaries for controlling accessibility and usage on the premises (Acosta and Palaoag, 2024; Purnama, 2024).

These are tools meant to actually boost security and have operations flow smoothly. However, instructors, students, and staff treat them as invasive. These parties are worried that somebody is watching them from every now and then, their information is not in safe hands, or leakage on cybersecurity breaches that pops up at one's sensitive information. This concern is heightened due to the fact that there isn't any clear policy in regard to data governance. There are likely to be issues of who owns the data, how it is used, or whether people can opt out from such surveillance without restriction. Beyond pragmatic issues, of course, there is a more profound cultural level on which unease that smart technologies in particular may eventually take universities further away from their traditional identity and values. Conventionally, colleges and universities have been regarded as places one goes to be with people, ideas, and thought processes. Critics say such institutions are in danger of losing the 'human touch' through increasing digitalization, becoming impersonal, data-driven, and algorithmically controlled. The increased preference for digital help among students, the instructors argue, demoralizes them in teaching in the classroom and damages the interpersonal relationship with their students. Alienation also strikes students when virtual barriers of communication hamper the personal time they are entitled to spend with peers and mentors. It is taking that sense of community away from students who see higher education as more than a transactional deal that defines campus culture. What this simply means is that the staff and instructors in these colleges are afraid that the smart campus has something to do with efficiency at the loss of tradition, focusing on getting things done in less time without necessarily maintaining the rituals, customs, and cultural preservation that make a place worth staying in. This is further increased by the fear that growing socio-economic divides will not be stemmed, since enumerated students from poorer backgrounds cannot afford full-scale participation in technologically driven education, adding to inequalities and making college less inclusive (Kumar et al., 2024; Chiang et al., 2024).

It should address Smart Campus Concerns as a strategic, inclusive process of priority through practices demanding transparency, collaboration, and equity. It is upon these bases that the colleges should install solutions to calm that fear by effecting change, which involves instructors, staff, and students through trusting relationships. Through the principle of paying attention to inclusive planning, open communication, and active participation by stakeholders, such learning enables these institutions to create an enabling environment where technological innovation upholds, instead of gnaws on traditional academic values (Keerthana, 2024).

Such objections can only be overcome with the help of participatory planning processes so that everyone's voice is equally represented in decision-making. Each college should be assigned a committee or advisory board of representatives of faculty, staff, students, and administrators. Each of these groups might have something very valuable to say about the possible implications of smart technologies for campus life in all its dimensions and allow the building of systems reflecting the interests and priorities of the whole community. Inclusive planning – questionnaires, focus groups, and forums – can be of help in understanding the views and apprehensions of the people against the implementation of new technologies. Shared ownership in this initiative is built in a way to reduce resistance by sharing in all aspects (Hong et al., 2024).

The openness of communication plays its part in dissipating the fear of job loss, privacy, and erosion of the culture. On the institutional side, transparency should be there regarding what purpose, scope, and implementation method are for the technologies of smart campuses. In fact, clear descriptions of how data are to be collected, stored and protected can allay concerns of privacy especially if the colleges make it known that they will follow legal standards and ethical practices. Periodic updates and progress reports reassure the stakeholders that their voice is still taken into consideration in the process. Beyond that, colleges have to provide opportunities where stakeholders can raise issues they would question, observe, or point their finger at on safe and supportive ground. That will help cement the relationships and underline fully that the institution values collaboration and accountability (Lu et al., 2024).

Of equal importance is stakeholder engagement in ensuring smart campus initiatives advance values and expectations that support the academic community, but could also make the instructors, staff, and students be actively a part of pilot programs and trials firsthand if scaling up was being done with newer technologies. Apart from the fact that it allows one to acquire firsthand hands-on experience with the new system to enable constructive critique of newer systems which one is going to operate with confidence. Additionally, this more professional avenue of training development program offers an institution some definite opportunities that an institution can show for change transitions in its staffs/ faculties technology. Instead, empower staff with the necessary competencies to function within a technology-based environment. This would diffuse the fear related to loss of jobs, and also place the colleges in a better position to make the workforce aware of new opportunities. However, with respect to erosion of culture, the colleges have got to consciously take steps to preserve values and social fabrics defining the nature of academic life. This might be modulated by weighing technology application against more human-centred approaches where smart systems are augmentative, not substitutional to interpersonal interaction. One such example could be that in one of these hybrid models, face-to-face teaching and the use of digital tools are integrated, hence offering flexibility but without losing the human touch that defines education (Sobol et al., 2024; Farhan and Daghal, 2024).

In-campus cultural activities involving celebration of cultural traditions, ceremonies, and rituals, symbolic acts, institutions, sense of belonging, and the identities thereof add meaning to the everyday campus lifestyle of all personnel that happen to dwell therein. It will ensure the digital equity policy is followed by each higher educational institution to see all the students benefit from smart campus initiatives irrespective of socio-economic statuses of any (Jonek-Kowalska and Wolniak, 2023; Jonek-Kowalska and Wolniak, 2025). The policies undertaken on subsidized technology programs may provide ways of renting devices at reasonable costs hence increasing access to the Internet and bridging up gaps in the digital divide. Instead, institutions should invest sufficiently in the different necessary programs for digital literacy among the students that will ensure competencies for handling the smart technologies. That would be equity and commitment to providing inclusive supportive learning (Polin et al., 2024; Bitencourt et al., 2024).

These are questions of privacy that would be better answered through strong data governance frameworks focused on security and transparency. Colleges should have formal guidelines over data collection, storage, and usage, availing options to opt-out when available to stakeholders (Kurniawan et al., 2025). Such credibility is enhanced through independent audits and third-party assessments that further reassure the stakeholder that data shall be handled responsibly (Table 9.3).

9.3 Identifying and Leveraging Opportunities

On-campus creativity has little to do with an impressive presentation of technology; instead, it evidences how learning environments will be created to enable collaboration, critical thinking, and the free flow of ideas. This will be possible by developing an interactive digital platform inclusive of interaction tools for presenting content to students in novel exciting ways, which will provide the ground for more personalized experiences in learning. Fair enough, AI, VR, and adaptive learning technologies have made learning personalized. As such, a whole array of learning opportunities can be gained that lie beyond the traditional confines of a classroom. In addition, abundant information and analytics in these platforms give instructors an array of options with regard to curricular activities in ways that intellectually challenge students to ensure creative thinking in solving some real-world problems. What these places contribute to is often grounds for this innovation: a place such as a lab space, an open

Table 9.3 Potential solutions to deal with cultural and social resistance in smart campus creation process

Solution	*Description*
Inclusive planning procedures	This will be done through committees and advisory bodies that are representative of faculty, staff, and students and will listen to every voice regarding decision-making. It seeks constructive feedback proactively through surveys, focus groups, forums on concerns.
Open communication	Also, communicate objectives, scope, and implementation of smart technologies. Provide information on updates, policies with regard to data use in simple terms, while simultaneously providing a podium for discussion in order to allay apprehensions and build trust among people.
Stakeholder participation	Engage instructors, staff, and students in pilots to test, assess, and refine new technologies prior to wide-scale adoption. Beyond that, training programs need to be planned to get comfortable with new systems for both the staff and faculty.
Digital equity and inclusion	Equitable Access Program: Very low rental costs for devices, low cost for internet access, and workshops on digital literacy create a uniform learning environment.
Preserving academic culture	To Do: Embrace technology and people-centred approaches that build in the human touch and the preservation of tradition. Leverage the hybrid learning model: Cultural traditions are to be celebrated within name and identity to the campus culture.
Data privacy and security policies	Develop appropriate structures and policies with regard to data acquisition, storage, and its application accordingly. Conduct independent audits and assessment activities related to compliance with the acts in regulations on security and privacy; address concerns of the stakeholders.
Job security and retraining programs	Retrain and upskill employees for newer technologies, emphasizing how these would enhance their roles, not replace them, to show commitment to workforce development tied with stability.

Source: Author's own work.

area, or collaboration space; virtual communities enable the students too to try stuff, experiment, but most importantly learn by mistakes, all ingredients of bringing out creativity (Chen et al., 2024b; Hideki et al., 2024; Wolniak et al., 2024; Wolniak and Jonek-Kowalska, 2022).

In this respect, issues related to sustainability provide multifaceted pathways toward enabling the smart campus to act responsibly for the environment. By leveraging the potential of various energy-efficient technologies, smart infrastructures, and sustainable building practices, they might reduce the ecological footprint and serve as a living laboratory in the creation and testing of sustainability initiatives. Other efforts include smart grids, renewable energy sources, and green building certification in order to minimize the waste and consumption of energy with a view to have a sustainable campus ecosystem.

Besides, digitization of campus operations, such as paperless classrooms, online resources, and cloud services goes a long way in saving resources to endorse environmental sustainability. Apart from environmental sustainability, one might also note that one of the interests that a smart campus could state refers to social sustainability, accessibility, and inclusiveness through technologies. It would be an option to have different learning resources available for students, equal opportunities in education, and the development of social interaction among different groups (Picallo et al., 2024, Zhang, 2024).

Needless to say, with innovative infrastructure and integrated technology included in smart campuses, student achievements will go up. Equipped with real-time data, performance analytics, and learning pathways, students are better placed to keep tabs on progress and get targeted support in areas where they need extra help. In turn, this will make it easier for educators to spot early patterns of challenge and reach out to them so that no one ever falls behind. It also generally offers a lot of extra and academic supporting activities to prepare for meeting the needs diversified from students for an all-round development. All these do not restrict one's intellect but take into consideration emotional and social growth, as it has always been seen that a wholesome student fares much better academically as well. Digital collaboration platforms that also allow them to get mentors and networks so they could connect to and learn more through their communities and further build their networks (Zarpellon et al., 2024).

Smart campus projects would therefore be a great opportunity to align infrastructure, technology, and academic offering with larger strategic goals. It would provide an opportunity, not only to make the institution more attractive but at the very forefront of innovation, to change the landscape. 'Attractive' to students today is no longer four walls but one offering state-of-the-art technology, active learning areas, and a totally innovative curriculum. The smart campus, with a well-integrated digital system and an environment of collaboration, is certainly going to be quite fascinating for the students of today in the search for a dynamic and proactive academic experience. With AI, machine learning, and virtual reality, lifelong, and personalized learning pathways, colleges will be able to offer experiences that no other modality will be able to match for better academic outcomes but also in high-in-demand skills in the modern workforce: problem-solving, adaptability, and creativity. Thus, a state-of-the-art learning environment in terms of cooperation and innovation will contribute, alongside feedback based on data, to the enhanced students' experience at the universities, thus rendering the institutions more attractive for highly qualified students both in their country and worldwide (Xiong et al., 2024; Ucar et al., 2024).

Another area in which smart campuses can pride themselves on having a commanding advantage is in securing research funds. The kind of environment that these campuses offer-that is, a digitally innovative atmosphere with advanced research equipment-embracing state-of-the-art research in so many varied fields is capable of doing everything. Because the college integrates all systems and promotes collaboration among its researchers-data sharing and interdisciplinary approaches-they are far better positioned to seek large-scale research projects. On adding smart infrastructure and analytics, better management of the resources would facilitate such institutions to stream forward their research processes for raising the productive efficiency level (Imbar, 2024). It has more easily traced-ergo, better evaluated-results from this kind of research and thus makes the possible concretes resulting from your work easy to present before various interested donors, governments, or even the private sector. This will actually enable one to acquire certain types of funds to further expand some researched initiatives. By positioning colleges as hotbeds for innovation and technological advancement, it would mean that not only financial backing but high-level partnerships with industry leaders, governmental agencies, and other philanthropic organizations flow into them (Liu, 2024b; Apiratwarakul et al., 2024).

The other objective that could easily be facilitated by smart campus projects is improvement in international standing (Table 9.4). Particular universities are

Table 9.4 How colleges can utilize smart campus initiatives to meet their strategic objectives

Strategic objective	*How smart campus initiatives support the objective*	*Importance of collaborations, technology transfer, and entrepreneurship*
Attracting top students	These campuses are equipped with state-of-the-art digital tools, interactive learning environments, and even personalized pathways to improve academic experiences. Furthermore, AI, VR, and adaptive technologies can be integrated for exceptionally attractive customized education experiences among the highest achievers.	International collaboration and partnership with other institutions and industries offer opportunities to network and international experience to students. Entrepreneurship programs and innovation hubs with entrepreneurial activities make the campus more attractive for students.
Securing research funding	Smart campuses provide an efficient infrastructure for research through the use of state-of-the-art technologies that enable collaboration, sharing, and interdisciplinarity. Data analytics and performance tracking in smart campuses are ways in which higher education can show quantifiable research output when applying for funding.	This initiative of technology transfer allows commercializing the research whereas generation of funds from the private sectors or industry partners is prevalent in nature. Strong liaisons with industries and governments enable the universities to generate outside funding for various research projects.

(*Continued*)

Table 9.4 (Continued)

Strategic objective	*How smart campus initiatives support the objective*	*Importance of collaborations, technology transfer, and entrepreneurship*
Enhancing international standing	Advanced technology applied on smart campuses depicts a university committed to innovation, thus adding to its reputation on the global stage. Smart campuses provide the opportunity for internationalization of connectivity in academic exchanges, international research partnerships, and sharing of resources.	International collaborations with universities, research institutions, and industries raise the profile and status of a college to the global level. Agreements on technology transfer and inter-border entrepreneurship increase respectability in developing international prestige for the university.
Collaborations, technology transfer, and entrepreneurship	Smart campuses create collaboration among students, faculty members, and even industry experts through digital platforms and real-time communication. In addition, the incubators and accelerators of entrepreneurial ecosystems have to be present in order for the students and faculty to commercially translate their ideas and innovations.	These partnerships with industry and government organizations provide a means for universities to convert research into commercially viable products and services, thereby enhancing innovation capacity and financial sustainability.

Source: Author's own work.

competing globally for standing and reputation. Smart campuses offer one avenue through which institutions can show interest in leading in technology matters, sustainability, and excellence in academic matters, all of which are crucial in determining global standing. With innovative technology, universities come into focus for a global research network, educators, and students and differentiate themselves from the competition. Through global connectivity, smart campuses help spur international collaborations and partnerships around shared goals, which really enable an institution to collaborate on an international scale with renowned experts, researchers, and thinkers. This increased exposure and collaboration raise the profile of the university and present options for collaborative research programs, student exchange across borders, and other jointly shared academic resources. Thus, institutions that adopt this model of a smart campus enjoy better international visibility with ease and probably contribute to shaping the global academic landscape (Budiono and Budiarto, 2024, Izourane et al., 2024).

References

Abulibdeh, A. (2024). Towards zero-carbon, resilient, and community-integrated smart schools and campuses: A review. *World Development Sustainability*, *5*, 100193.

Acosta, K. P., & Palaoag, T. D. (2024). Design of a gender responsive smart campus surveillance with iot integration and stakeholders' adaptability. *Proceedings on Engineering Sciences*, *6*(4), pp. 1465–1472.

Alvarez, C. I., Armijos, E., Morales, G. S., Aguilar, K., & Cadena, J. (2024). Geospatial digital twins for the development of smart sustainable campus initiatives in Ecuadorian Catholic Universities. In *IEEE Andescon, ANDESCON 2024 – proceedings*.

Apiratwarakul, K., Cheung, L. W., Pearkao, C., Gaysonsiri, D., & Ienghong, K. (2024). Smart emergency call point enhancing emergency medical services on University Campuses. *Prehospital and Disaster Medicine*, *39*(1), 32–36.

Baba, K., Elfaddouli, N.-E., & Cheimanoff, N. (2024). The role of information and communication technologies in developing a smart campus with its four pillars' architectural sketch. *Education and Information Technologies*, *29*(12), 14815–14833.

Bandória, L. H. T., Cortes, B., & de Almeida, M. C. (2024). Statistical characterization of electricity use profile: Leveraging data analytics for stochastic simulation in a smart campus. *Energy and Buildings*, *324*, 114934.

Bellaj, M., Bendahmane, A., Younes, A., Boudra, S., & Ennakra, M. (2024). A systematic review of smart campus technologies to improve students' educational experiences. In *Proceedings of 2024 1st edition of the Mediterranean smart cities conference, MSCC 2024*.

Bitencourt, L., Borba, B., Dias, D., Lopes, Y., & Fernandes, N. (2024). Shared e-bike operational planning under different relocation schemes: A smart campus case. *Results in Engineering*, *22*, 102368.

Budiono, F. L., & Budiarto, H. (2024). The implementation of smart campus design for the Jogja digital polytechnic. In *11th International conference on ICT for smart society: Integrating data and artificial intelligence for a resilient and sustainable future living, ICISS 2024 – proceeding*.

Chagnon-Lessard, N., Gosselin, L., Barnabe, S., Vandersickel, A., & Zhang, P. (2021). Smart campuses: Extensive review of the last decade of research and current challenges. *IEEE Access*, 9, 124200–124234.

Chen, C., Yang, Y., Yuan, H., Qu, B., & Yang, D. (2024a). Animating the crowd mirage: A WiFi-positioning-based crowd mobility digital twin for smart campuses. *Proceedings of the ACM on Interactive, Mobile, Wearable and Ubiquitous Technologies*, *8*(4), 158.

Chen, H., Lin, W., & Lai, C. (2024b). Research on decision optimization for smart campus in colleges and universities based on feedback mechanism to improve data quality. In *Lecture notes in electrical engineering, 1256 LNEE* (pp. 877–884).

Chiang, L.-C., Lin, W.-T., Tai, K.-T., Li, D.-C., & Wu, P.-J. (2024). The Assessment of generation Z's acceptance of technology-enhanced public bicycle-sharing system in the construction of smart mobility through student-centricity solutions in NCKU campus in response to tainan urban district railway underground project. In *International conference on eDemocracy and eGovernment, ICEDEG, 2024*.

Das, D., Lim, N. D., & Aravind, P. (2022). Developing a smart and sustainable campus in Singapore. *Sustainability*, *14*(21), 14472. https://doi.org/10.3390/su142114472 (Accessed 17.01.2025).

Delnevo, G., Ghini, V., Fiumana, E., & Mirri, S. (2024). A support tool for emergency management in smart campuses: Reference architecture and enhanced web user interfaces. *Sensors*, *24*(18), 5887.

Domínguez-Bolaño, T., Barral, V., Escudero, C. J., & García-Naya, J. A. (2024). An IoT system for a smart campus: Challenges and solutions illustrated over several real-world use cases. *Internet of Things*, *25*, 101099.

dos Santos Sá, J., Do Nascimento Da Silva, E., Nunes Gonçalves, L., José Brito Barros, F., & de Souza Farias, F. (2024). Multimodal urban mobility solutions for a smart campus using artificial neural networks for route determination and an algorithm for arrival time prediction. *Engineering Applications of Artificial Intelligence, 137*, 109074.

Eindhoven University of Technology Expands LoRaWAN® Connectivity with TEKTELIC Gateways. (2024). https://tektelic.com/news/eindhoven-university-and-tektelic-enhance-iot-smart-building-solutions-with-kona-enterprise-gateway/ (Accessed 17.01.2025).

Eldeeb, E., & Alves, H. (2024). LoRaWAN-enabled smart campus: The data set and a people counter use case. *IEEE Internet of Things Journal, 11*(5), pp. 8569–8577.

Famoriji, O. J., & Shongwe, T. (2024). Simple and effective electromagnetic wave propagation loss model in GSM band for smart campus applications. *SSRG International Journal of Electrical and Electronics Engineering, 11*(11), 306–311.

Farhan, H. L., & Daghal, A. S. (2024). Sustainable smart campus by assisting IoT. *AIP Conference Proceedings, 3092*(1), 040026.

Florentin, K. M., Onuki, M., & Yarime, M. (2024). Facilitating citizen participation in greenfield smart city development: The case of a human-centered approach in Kashiwanoha international campus town. *Telematics and Informatics Reports, 15*, 100154.

Geng, J., Tang, B., Zhang, B., Shao, J., & Luo, B. (2024). Demo: FedCampus: A real-world privacy-preserving mobile application for smart campus via federated learning & analytics. In *Proceedings of the international symposium on Mobile Ad Hoc Networking and Computing (MobiHoc)* (pp. 377–378).

Hassanain, M. A., Sanni-Anibire, M. O., & Mahmoud, A. S. (2024). An assessment of users' satisfaction with a smart building on university campus through post-occupancy evaluation. *Journal of Engineering, Design and Technology, 22*(4), 1119–1135.

Hideki Mensch Maruyama, M., Willig Silveira, L., Da Silva Júnior, E., Palazzo, M. De Oliveira, J., & Maran, V. (2024). Recommender systems in smart campus: A systematic mapping. *Knowledge and Information Systems, 67*(3), 2063–2089.

Hong, W., Huang, Z., Zhang, J., Chen, X., & Chen, B. M. (2024). Pathway to a smart campus framework: A review and case study. In *IEEE international conference on control and automation, ICCA* (pp. 424–430).

Hussien, S. H. T., Farhana, H. A. R. N., Vinukumar, L., Alexander, C. H. C., & Sivakumar, S. (2024). Smart campus attendance and security systems: Leveraging IoT for enhanced efficiency. *AIP Conference Proceedings, 3161*(1), 020153.

Imbar, R. V. (2024). Approaches that contribute to a new definition of the 'Smart Campus' based on smart systems. In *11th International conference on ICT for smart society: Integrating data and artificial intelligence for a resilient and sustainable future living, ICISS 2024 – proceeding.*

Izourane, F.-Z., Ardchir, S., Ounacer, S., & Azzouazi, M. (2024). Smart campus based on AI and IoT in the era of Industry 5.0: Challenges and opportunities. *Studies in Systems, Decision and Control, 565*, 39–57.

Jonek-Kowalska, I., & Wolniak, R. (2023). *Towards sustainability and a better quality of life?* London: Routledge.

Jonek-Kowalska, I., & Wolniak, R. (2025). *Exclusion in smart cities assessment and strategy for strengthening resilience*. London: Routledge.

Keerthana, P., Prasath, S., Tamilmani, S., & Veeragokulraj, S. (2024). Enhancing campus security through smart surveillance system. In *10th International conference on advanced computing and communication systems, ICACCS 2024* (pp. 876–881).

Kim, I., & Lee, A. J. (2024). I know what you did last semester: Understanding privacy expectations and preferences in the smart campus. In *Conference on human factors in computing systems - proceedings* (p. 980).

Kumar, B., Padmanabha, V., Josephmani, P., & Dwivedi, R. (2024). Utilizing packet tracer simulation to modern smart campus initiatives for a sustainable future. In *2024 IEEE 12th international conference on smart energy grid engineering, SEGE 2024* (pp. 97–102).

Kurniawan, F., Aziza, M. R., Hasanah, N.A., Wibawa, A. P., & Hammad, J. (2025). The innovative smart green campus as life-based learning characteristics of future learning efforts to complete the SGD's, *Journal of Lifestyle and SDG'S Review*, *5*(2), e02908.

Li, Y., Chen, H., & Yu, P. (2024). Green campus transformation in smart city development: A study on low-carbon and energy-saving design for the renovation of school buildings. *Smart Cities*, *7*(5), 2940–2965.

Liang, J., & Gong, Y. (2025). Security and privacy protection protocol based on edge computing in smart campus. *Expert Systems*, *42*(1), e13579.

Lin, Z., Liu, W., Fu, X., & Lian, B. (2024). A method for analyzing and warning the safety behavior of smart campus personnel based on deep learning. *Frontiers in Artificial Intelligence and Applications*, *382*, 1111–1118.

Liu, X. (2024a). Research on Rapid parameterized modeling of cityengine in the application of smart campus platform. *E3S Web of Conferences*, *512*, 01027.

Liu, Z. (2024b). Research and application of behavior analysis of college students based on smart campus. In *ACM international conference proceeding series* (pp. 457–464).

Lu, Y., Hu, X., & Li, J. (2024). Smart campus: the deep integration of machine vision and physical education. *Transactions of the Royal Institution of Naval Architects Part A: International Journal of Maritime Engineering*, 1(1), A163–A173.

Mariappan, R., Suhasini, P. S., Lakshmi, G. M.,Sai, M. M. S., & Devendhar, R. (2024). Data visualization for energy optimisation in smart campus application using LoRaWAN industrial IoT technology. *AIP Conference Proceedings*, *3031*(1), 020004.

Molenaar, T. (2020). *A smart campus as a test case for the users' perception of location data in smart cities*. https://studenttheses.uu.nl/bitstream/handle/20.500.12932/39140/GIMA%20Final%20Thesis%20Report%20T.%20Molenaar.pdf?sequence=1 (Accessed 17.01.2025).

Pexyean, T., Saraubon, K., & Nilsook, P. (2024). Digital twin energy management system with artificial intelligence internet of things to smart campus. *Journal of Theoretical and Applied Information Technology*, *102*(17), 6351–6363.

Picallo, I., Klaina, H., Lopez-Iturri, P., Villadangos, J., & Falcone, F. (2024). Development of a cognitive IoT-Enabled smart campus. In *2024 14th International symposium on communication systems, networks and digital signal processing, CSNDSP 2024* (pp. 644–647).

Polin, K., Yigitcanlar, T., Washington, T., & Limb, M. (2024). Unpacking smart campus assessment: Developing a framework via narrative literature review. *Sustainability*, *16*(6), 2494. https://doi.org/10.3390/su16062494

Purnama Alamsyah, D., Morika, D., Siswanto, B., & Othman, N. A. (2024). Smart campus and education: The study of individual performance in e-learning environments. In *11th International conference on ICT for smart society: Integrating data and artificial intelligence for a resilient and sustainable future living, ICISS 2024 – proceeding*.

Report of Quality Audit of the Chinese University of Hong Kong. (2024). *Quality assurance council third audit cycle*. https://www.ugc.edu.hk/doc/eng/qac/report/cuhk202406e.pdf, (Accessed 17.01.2025).

Roda-Sanchez, L., Cirillo, F., Solmaz, G., Olivares, T., Kovacs, E. (2024). Building a smart campus digital twin: System, analytics, and lessons learned from a real-world project, *IEEE Internet of Things Journal*, *11*(3), 4614–4627.

Smart campus development in the greater Bay Area: Working towards a sustainable future (2024). https://www.ruijienetworks.com/about/news/greaterbayarea (Accessed 17.01.2025).

Smart campus (2025). https://www.ntu.edu.sg/about-us/smart-campus (Accessed 17.01.2025).

Smart campus, smarter thinking. (2023). Nanyang Technological University, Singapore, https://www.ntu.edu.sg/docs/default-source/corporate-ntu/ntu-corp-brochure_210609_(reading).pdf?sfvrsn=e4a2558e_2 (Accessed 17.01.2025).

Sobol, A., Kochyn, V., Huk, A. (2024). Information security system models in smart campuses. In *Proceedings - 2024 international conference on industrial engineering, applications and manufacturing, ICIEAM 2024* (pp. 750–754).

Stecuła, K., & Wolniak, R. (2022). Advantages and disadvantages of e-learning innovations during COVID-19 pandemic in higher education in Poland. *Journal of Open Innovation: Technology, Market, and Complexity*, *8*(3), 159.

Sustainability plan 2030. (2024). https://about.unimelb.edu.au/__data/assets/pdf_file/0020/346214/Sustainability-Plan-2030.pdf (Accessed 17.01.2025).

The Road to Intelligent Education Networks: The University of Hong Kong Partners with Huawei to Build a Wi-Fi 7 High-Quality Smart Campus (2024). https://e.huawei.com/at/case-studies/solutions/enterprise-network/2024-hong-kong-university (Accessed 17.01.2025).

Ucar, C., Maloku, M., Yugay, O., & Budimir, D. (2024). IoT motion detection sensors for monitoring in a smart campus. In *2024 13th mediterranean conference on embedded computing, MECO 2024*.

Valks, B., Arkesteijn, M. H., Koutamanis, A., & den Heijer, A. C. (2020). Towards a smart campus: Supporting campus decisions with Internet of Things applications. *Building Research & Information*, *49*(1), 1–20.

Venkatesan, M., & Aatmaja, A., & Mahariba, A. J. (2024). IoT based smart and efficient waste collection system for college campuses. *AIP Conference Proceedings*, *3075*(1), 020022.

Wang, K.-C., Pan, H.-W., & Wu, C.-E. (2024). Smart campus innovative learning model for social practitioners of universities' third mission: to promote good health and well-being. *Sustainability*, *16*(14), 6017.

Wolniak, R. (2023). Analysis of the bicycle roads system as an element of a smart mobility on the example of Poland Provinces. *Smart Cities*, *6*(1), 368–391. https://doi.org/10.3390/smartcities6010018

Wolniak, R., & Stecuła, K. (2024). Evaluation of quality of innovative e-learning in higher education: An insight from Poland. *Applied System Innovation*, *7*(6), 109. https://doi.org/10.3390/asi7060109

Wolniak, R., Gajdzik, B., Grebski, M., Danel, R., & Grebski, W. W. (2024). Business models used in smart cities—theoretical approach with examples of smart cities. *Smart Cities*, *7*, 1626–1669.

Wolniak, R., & Jonek-Kowalska, I. (2022). The creative services sector in Polish cities. *Journal of Open Innovation: Technology, Market, and Complexity*, *8*, 17.

Xiong, X., Liang, J., & Hu, J. (2024). Research and design of smart campus security information recognition and architecture based on LoRaWAN. *Proceedings of SPIE - The International Society for Optical Engineering*, *13185*, 1318500.

Xu, X., & Zhao, X. (2024). Exploration of the construction path of smart campuses integrated with Artificial IntelliGence-generated content. In *2024 7th International conference on computer information science and application technology, CISAT 2024* (pp. 1292–1295).

Yin, M., Li, K., & Huang, Z. (2024). Exploring sustainable design of smart campus restorative landscapes - - based on GIS Two-step floating catchment area method. In *ACM international conference proceeding series* (pp. 65–69).

Yu, H., Zhang, Q. (2024). Research on psychological screening and early warning for smart campus based on privacy protection. In *ACM international conference proceeding series* (pp. 685–690).

Zarpellon, B. O., De Oro Arenas, L., Paciencia Godoy, E., Pinhabel Marafao, F., & Morales Paredes, H. K. (2024). Design and implementation of a smart campus flexible internet of things architecture on a Brazilian University. *IEEE Access*, *12*, 113705–113725.

Zhang, Y. (2024). Intelligent smart campus management system based on full-time video surveillance platform. In *2nd International conference on sustainable computing and smart systems, ICSCSS 2024 - proceedings* (pp. 1331–1335).

10 Policy, Governance, and Regulation in Smart University Integration

10.1 National and International Policies Supporting Smart Universities

Most countries, in turn, finance it with a national education budget, research grants, and innovation funds with focuses on technology to create smart campuses. Similarly, many financial schemes have been developed in such a way that equitably shares the benefits in rural and urban areas accruing from digital learning tools and infrastructure. This is not all, the funding goes right down to training programs for teachers, research collaboration between universities and technology companies, and the development of digital learning content. Equally often, governments go in partnership with international organizations with the view to attracting funds from global initiatives that seek to raise the profile of higher education through the use of digital technologies (Nguyen and Nguyen-Dinh, 2024).

A very good example is how the Digital Education Action Plan of the European Union (EU) lays down a guideline in respect of developing digital learning for the entire EU, thus assisting nations. The EU, under initiatives such as the Erasmus+ program, has supported multiple projects with the aim of getting modern universities interested in developing these tools and international collaborations for Smart Education. The government of countries like South Korea and Singapore has involved itself in a holistic strategy of making the university the centre of innovation and technology-driven learning with high emphasis on research and development, apart from creation of a regulatory framework that allows smooth integration of digital technologies (Digital Education, 2020).

In turn, the regulatory framework plays a major role in aligning smart universities with societal needs and ethical considerations against data protection standards. Different national policies take up certain legislation that defines such cases related to intellectual property rights, cybersecurity, data privacy, and ethical usage of artificial intelligence in educational organizations. For example, the EU has already set up the GDPR (General Data Protection Regulation) to determine a more stringent legal framework on student and staff privacy, but also with full disclosure of how their personal information is

DOI: 10.4324/9781003621829-11

processed. Many other governments in their own way have similarly shown increased interest in policy development that balances the need for digital innovation with the protection of individual rights (Legal framework, 2018.

In this regard, there has been collaboration and agreements globally to allow for the advancement of smart universities. International organizations, like UNESCO and the World Bank, also provide a venue for knowledge sharing and dissemination of best practices in digitized education. With international partnership, the university will be able to get different types of expertise, technology solutions, and finance. For instance, the SDG (Sustainable Development Goals) 4 of the UN – for ensuring inclusive and equitable quality education – has been one of the key drivers for initiatives that have been driving universities to integrate smart technologies into teaching methodologies (Goal 4, 2020).

But as globalization keeps framing higher education, supportive policies for cross-border education are the next best thing. The international collaboration being fostered in universities and technology providers promotes the creation of a global network of smart campuses where innovative solutions, digital curricula, and research projects will be shared. Most collaborations are supported through various bilateral and multilateral agreements that have emphasized the role of digital transformation in education for economic growth and social progress (Pinto, 2024; Wolniak and Jonek-Kowalska, 2022).

Other guiding materials concern issues like sustainability and international competitiveness of smart universities. Legal frameworks developed both nationally and internationally do provide an environment that promotes the support of innovative technologies and develops a long-term position concerning sustainability that will ensure that higher education stands globally in respect. Some of these laws range from data protection to environmental sustainability that have gone a long way in changing universities into smart, technologically adept institutions (Tomas-Miquel et al., 2023; Al-Jayyousi and Aldhmour, 2023).

The very supportive role of the laws regarding smart technologies in universities mostly involves creating a very transparent regulatory framework that would ease the integration of digital innovations. Several countries have enacted legislation to integrate such technologies as artificial intelligence, machine learning, the Internet of Things, and big data analytics into educational institutions. These legislative frameworks encourage innovation in universities through tax breaks or grants for technology upgrades, or by creating a regulatory environment that allows for collaboration between higher education and technology companies. Tax credits can be provided for smart infrastructure investments at universities in energy-efficient buildings or to adopt high-level digital platforms that help enhance learning and administrative efficiency. These types of incentives make the transition to smart technologies economically feasible while encouraging innovation within the industry (Salomaa et al., 2023; Pirlone et al., 2020).

On the other hand, legislation also covers the contribution of universities to sustainable development through technology. Different governments have, therefore, launched legislation that requires universities to implement their contributions to the commitment to sustainability at the national and international levels. The EU's Green Deal and the United Nations' SDGs provide a lawful and policy-driven impetus for incorporating sustainability at universities. Over the years, many countries have enacted legislation that demands more involvement from universities in reducing carbon emission, energy consumption, and waste. These goals are significantly related to smart technologies as they enable universities to monitor and manage the usage of resources effectively. First, energy management systems can be installed smartly to optimize electricity and water consumption. IoT devices may monitor waste generation and recycling processes (Tomasi et al., 2024). Legal stimuli may pertain more specifically to taking up renewable sources of energy, such as solar or wind, coupled with intelligent distribution that improves energy sharing. Legal frameworks are very relevant drivers to enable the uptake of the CE by higher educational institutions. Legislation in the area of waste management and recycling issues furnishes parameters within universities with respect to how they are implementing sustainability in their operations or shrinking their ecological footprint. In fact, smart technologies have made this possible, especially at those universities that apply data analytics towards further optimization of procurement, the reduction of consumption, and, finally, better lifecycle product management. In fact, the legal frameworks that are for sustainable innovation play an important role in ensuring that universities not only adopt smart technologies but are able to use them to add positively towards environmental aspects (Dodgson and Staggs, 2012; Spanoudakis et al., 2024).

Laws also provide a footing on which universities can strategically apply technology to develop their international competitiveness. Meanwhile, international agreements and national policies foster cross-border research collaboration, digital education, and the sharing of technological resources to improve the global outreach of universities. In this respect, the legal frameworks create avenues for collaboration with international universities and technology firms, enabling smart universities to access global expertise and resources. It does so, for one, by promoting international academic exchange with the help of laws and regulations that guarantee protection of intellectual property rights, taking part in international research projects, or transferring technologically relevant findings from abroad (Tse et al., 2021). On the other hand, digital learning platforms or virtual exchanges have thus also created incentives towards the internationalization of higher education. In fact, in many cases, legal frameworks have been developed to break barriers in cross-border education through limitations in online learning and international standards of accreditation. As the concept of smart universities progresses to either hybrid or completely digital learning environments, such supportive legal frameworks will finally allow higher learning institutions to include more international students and collaborate more widely with other institutions. All these activities make universities not only internationally more competitive but also motivated in curricula creation relevant for real labour markets worldwide shaped in the new context of digital economics (Zagrajek et al., 2023; Lamey et al., 2023).

With that in mind, the most strict laws on Data Security and Privacy – a product of general European legislation concerning GDPR – gain significant importance as drivers of using modern technologies at universities. These laws strike a balance in protecting the students' data and in building confidence in digital systems used for educational purposes. Stringency of data protection laws provides credibility to the universities as a way of attracting international partnerships and students. While universities move on to increasingly sophisticated technology-driven learning and methodologies for research, including but not limited to AI-powered personalized learning systems, these legal protections need to be robust for the assurance of ethical usage of the data (Mircea et al., 2021).

Global organizations and international partnerships also have a very important role in shaping guidelines and advocating for the exchange of best practices with regard to developing smart campuses. Global organizations provide ways in which knowledge, resources, and expertise are shared in the globally increasingly interconnected higher education setting, thus fast-tracking smart technology adoption in educational settings. Partially, because global organizations do drive the development of smart campus in some ways: by developing frameworks, creating partnerships, building collaboration networks, thus giving universities a chance to keep up with rapid technology advancement against the many different challenges the fast-moving world places on education. Other key roles played by global organizations in respect of this are setting international guidelines and standards that guide the deployment of smart technologies on campus (Pirlone et al., 2020; Dodgson and Staggs, 2012). Other organizations, such as UNESCO, the United Nations, and the World Bank, have framed digital technologies as transformative for education and provided frameworks that call for their responsible integration. For instance, UNESCO's 'Digital Education Strategy' laid down the guidelines for adopting digital technologies in educational institutions, even to an extent that the policies related to smart campuses ensured that there was access, equity, and inclusion. In turn, this creates a set of shared understanding among countries and institutions towards technology-driven education and guiding foundational principles on smart campus development in accessibility, data privacy, and sustainability (Kangas et al., 2020; Zagrajek et al., 2023; Tomasi et al., 2024).

While the contribution has been highly fragmented, international organizations have at least tried to harmonize such works across countries and institutions for cooperation to try and avoid fragmentation in approaches towards the development of smart campuses. Though many universities can be found to work in complete isolation, global collaborations assure shared learning and solving together problems. Therein, an international network and partnership will be built for the development of best practice, technological solution, and lesson learned exchange. These may be partnerships where universities link their resources, advise on matters such as the use of digital tools to improve learning environments – for instance, the Global Partnership for Education. Actually, most of those global initiatives brought together the university and government, technology providers, and non-governmental organizations all as one comprehensive supportive system towards innovation and digital transformation in higher education (Santos, 2019; Pinto, 2024).

It also made further the aspect of international cooperation to facilitate cooperation in sharing policies and best practices, hence providing a means that could be taken in helping pool resources, expertise, and technological innovations that are required by universities for their development into smart campuses. It is actually this type of bilateral and multilateral agreement that even allows universities to partner with leading technology companies and research institutions in the application of state-of-the-art innovations in campus development. All these come into play as a partnership between institutes of higher learning with technology companies in using the most appropriate intelligent solutions with respect to energy-efficient buildings, modern state-of-the-art learning management systems, and AI-powered learning tools. It sometimes encompasses joint research into new technology and solutions which could just be at an experimental state of their being and thereby a game changer for Higher education (Tomas-Miquel et al., 2023).

International organizations also align the development of smart campuses with international standards in view of broader goals, such as sustainability and social inclusion. It then goes ahead to provide a guide for the development of smart campuses, especially on issues regarding ensuring inclusive and equitable quality education based on United Nations' SDGs, notably Goal 4. Global organizations call for the adoption of smart technologies in realizing sustainable goals through universities by encouraging campus designs integrated with renewable energy sources, waste management systems, and energy-efficient infrastructure. Global networking also helps such organizations diffuse best practices towards building smart campuses that are not only technologically advanced but equally environmentally sustainable and socially responsible (Salomaa et al., 2023; Al-Jayyousi and Aldhmour, 2023).

Global partnerships in higher education institutionalize internationalization and enable the establishment of university campuses that could attract students and staff from across the world. Equipped with intelligent campus capability, digital tools, and online learning platforms, smart campuses would be able to extend flexible learning opportunities across borders for the benefit of a global audience. This would also mean international collaboration, supported by international organizations, enabling universities to offer cross-border programs and virtual exchanges, adding to the global competitiveness of Smart Campuses. This would hence make them remain relevant in the increasingly internationalized higher education sector, ahead in innovation, and ultimately truly diverse and global learning experience for their students (Tomasi et al., 2024).

10.2 University Governance Models for the Digital Age

These governance models relate to effective management and implementation that the colleges may follow in governing the smart campus projects. Such are multi-faceted, comprising a variety of roles and processes in an integrated manner to ensure smart campus initiatives realize their goals. Of these, stakeholder engagement is one of the most important elements in a combination of interests and needs from the different viewpoints of groups either involved or affected by the project. Engagement builds ownership into the stakeholders; it fosters cooperation and ensures that all relevant perspectives are considered, both at the planning level and at the level of implementation. This would presumably include students, faculty, staff, and administrators along with technology and service providers, and outsiders ranging from government representatives to industry experts (Ahriz et al., 2024; Bassano et al., 2019; Wolniak, 2023; Wolniak et al., 2024).

The leadership within the governance framework is very important. As a matter of fact, for the smart campus projects to be led in the right direction towards successful completion, strong and visionary leadership is called for. Their decisions have to be taken by balancing innovation with practicality, deeply cognizant both of the technological and educational perspectives that surround the project. The leaders should guarantee a trusting, transparent, responsible atmosphere. They have to communicate effectively with all stakeholders; drive the project through the hard conditions – mostly high in complexity and often of disruptive nature – when implementing smart campus technologies. This here means identifying, mitigating, and navigating around risks and challenges effectively; adapting or changing plans where necessary. A decision will always keep the project leading, heading into the direction of the realization of the project's overall objectives from an institutional perspective. However, the framework for the decisions made has to be adequately fast and agile enough to institute immediate changes while being firm on what is laid to protect it from going off their course or focus set of the work (Man et al., 2023; Abdulrasool and Turnbull, 2020; Jonek-Kowalska and Wolniak, 2023; Jonek-Kowalska and Wolniak, 2025).

Decision-making mechanisms in Smart Campus projects often involve several layers, with various consultations at various levels. This forms part of a very integral process whereby clear criteria have to be put in place against which options can be evaluated and selected based on data. The necessity of taking decisions with long-term sustainability relevance, but not just with a regard for short-term technological needs, and the larger implications for campus life and academic performances, must be ensured (Soni et al., 2023). Success, on all these counts of the Smart Campus Project, therefore, hinges together: engagements, leadership, decision-making. Each of these aspects must be judiciously welded into one coherent governance model for the effective implementation of, and with regard to the changing times concerning technology and education, the strategic vision of the institution. It could do so by turning this trend of initiatives on the development of smart campuses towards a far-enriching scope: to enhance the learning environment, operational efficiency, and even the integral part to play in integrative development for sustainable and innovative education (Chen and Liu, 2021; Mario et al., 2021; Wolniak and Stecuła, 2024; Stecuła and Wolniak, 2022).

Both technological developments and the needs of the campus community are themselves dynamic entities; governance frameworks need to be clear in structure yet able to accommodate change over time. Systems will be designed deeply cognizant that higher education, along with supportive technology, is in continuous movement. Otherwise, without the ability to adapt, governance would risk becoming rigid, outdated, or unable to act upon emergent challenges or opportunities (Lee and Miozzo, 2024).

Governance transparency in the area of higher education gives students, faculty members, and staff confidence and clarity as to how and why decisions have been made by whom (Guenduez et al., 2024). It does reduce uncertainty and allows for more involvement in decision-making, strengthening a culture of accountability within a university or college environment. This is the type of initiative that stakeholders – all of them informed as to the goals, processes, and results – will support out of a sense of ownership. Besides, transparency may also reveal probable conflicts of interest in due time and enable the alignment of the multiple standpoints, reducing misunderstandings or resistance during the implementation phase (Lian and Chen, 2024; Chagnon-Lessard et al., 2021).

While that may be so, the mere transparency of the processes cannot substitute for it; there also needs to ensure that such systems are 'agile' in catering to the dynamic needs coming from the campus community. All pointers – the diversity of perspectives, changeable character of the practices followed, and capabilities of a continuously improving technological base – carry an essential message towards what future change must reach within evolving governance structures together with shifted priorities and pressures (Ma, 2024). It is in this respect that one governance model remains relevant to the context in which it finds itself. In such a way, an agile approach to governance could place an institution in a position to pivot and bend in new ways towards the resolution of an unforeseen challenge, say of funding, shifting student needs, or technological disruption. Adaptability will ensure that the new technologies or processes absorbed by the governance system do not change the culture and therefore make the process less cumbersome and smooth for all participants of the process (Kim et al., 2024; Mario et al., 2021).

Governance structures provide that the smart campus initiative keeps fresh and adaptive, hence securing its continued relevance in serving the interests and needs of the community through times. In this way, continuous refinement processes could be allowed for infusing ideas through which the institution will put into facing the demands of an increasingly wiring and rapidly changing academic environment. This, in turn, could help to establish a sound governance structure in the interest of all stakeholders in collaborative effort on further over-arching goals of sustainability, accessibility, and excellence in education (Liu, 2020; Telino et al., 2020). Table 10.1 present the summarization of main difficulties in preparation of smart campus program.

Macau University of Science and Technology and Huawei can be a case study in their strategic partnership to construct such urgently needed facilities

Table 10.1 Difficulties in preparing smart campus programs

Difficulty	*Description*
Stakeholder coordination	It can be a recipe for communication breakdowns, conflicts of goals, and a mess in decision-making when there is multi-stakeholder management with various interests and priorities. Ensuring alignment and collaboration is key.
Resource allocation	Resources such as funding, personnel, and time distributed appropriately among the many dimensions of a smart campus may be complex to handle while there are competitive priorities by stakeholders.
Technology integration	All that's in place needs to be harmoniously joined with new technologies, adding their bits to infrastructure. Complexities include incompatibility and the threat of outdatedness.
Ensuring strategic alignment	Ensuring alignment between smart campus projects with long-term university goals; otherwise, the misfit of objectives might result in useless investment and solutions that might not work out effectively.
Managing change	The rapid pace of technological change and changing campus needs demand governance systems capable of rapid adjustment. An inability to adjust to new challenges or opportunities will inhibit progress.
Decision-making complexity	Decision-making for multi-party projects often requires reconciling multiple perspectives and expertise, and it is not always easy to achieve consensus. A clear decision-making structure and transparent processes are required.
Risk management	Identifying and mitigating risks – technological failure, budget overruns, dissatisfaction by stakeholders – is crucial for the success of the project. Development of proactive risk management strategies is called for.
Measuring impact	Tracking tangible smart campus outcomes, such as improved student learning, efficient operation of the campus, and sustainability, can be challenging. Clear metrics and periodic assessments are required.
Maintaining flexibility	As projects progress, it may be that the campus environment or external factors such as changes in policy or budgetary constraints will change; any governance system needs to allow flexibility in adjustments without compromising their original objectives.
Ensuring sustainability	There is a challenge in creating financial and operational sustainability of the long-term smart campus program. Institutions need to become forward-looking in planning for ongoing maintenance, updating, and scalability.

Source: Author's own work.

to build a smart campus, promoting culture, developing intellect, and propelling economic progress in tune with the 2021–2025 Strategic Plan of the university. The university governance will support this model in carrying out research innovation, intelligent teaching, management refinement, and proactive services. M.U.S.T. (Macau University of Science and Technology) also aims to develop high-end technologies for education quality improvement and managerial efficiency. For further details on the collaboration, visit the website of Huawei (CTM, 2023; The Digital, 2024; Macao, 2024).

Stanford University has developed a governance model in its transition to a smart campus that lays quite an emphasis on sustainability, operational efficiency, and data-driven decision-making. In the case of the Sustainable Stanford initiative, the suite of state-of-the-art technologies implemented measures energy usage and optimizes resource use, therefore attempting to reduce environmental impact. This is backed up by a decentralized yet coordinated governance framework where different actors, including faculty, administrators, and students, are involved in representing their respective interests. It is not out of place to mention that Stanford's Office of Sustainability is the major actor in the governance structure related to the planning and implementation of the projects dealing with the smart campus. Stanford has invested in smart energy systems, such as the Stanford Energy System Innovations program, that decrease greenhouse gas emissions by 68%, a real-time data collection and analytics platform to track energy usage, water efficiency, and waste management. Sustainability committees develop decision-making processes in such a manner that progress can be reviewed and recommendations for adjustment proposed because needs will continue to evolve on campus. Moreover, the Stanford governance model shows much in support of transparency through annual reports in sustainability and community involvement by piloting projects and conducting research collaborations with students and staff. This will make sure this model remains aligned with the long-term university goals while having elasticity to adapt to changes in technology or shifting priorities (Smart Campus, 2024; min-Allah and Alarashed, 2020; Snider, 2022).

The Hong Kong University of Science and Technology has adopted a pioneering, innovative framework of governance: the 'Sustainable Smart Campus as a Living Lab' program. Beginning in 2024, this program transforms the university into a testing ground where ideas are conceptualized, experimented with, and implemented for finding solutions on sustainability and operational issues. This model of governance is inclusive and collaborative in nature. It makes the solutions advanced yet socially inclusive and environmentally responsible by active participation of faculties, researchers, and students in a variety of decision-making aspects. HKUST (Hong Kong University of Science and Technology) established a central steering committee to oversee project planning and execution, thus providing complete transparency and accountability throughout the life cycle of the initiative. The HKUST Smart Campus framework puts into perspective priorities of integrating renewable energy, smart water management, and deployment of IoT devices in monitoring building performance. The governance structure provides ample space for creating partnerships with technology providers and government organizations that make the initiatives scalable and sustainable long into the future. Continuous monitoring and assessment of the program are carried out. Annual reviews are done, and impact assessments are carried out to make further changes. The Living Lab concept also works as an educational tool: it actively involves students and faculty through research opportunities in fulfilling the overall sustainability goals of the university (Sustainable, 2024; Pham et al., 2020; Polin et al., 2023).

10.3 Regulatory Challenges and Compliance Strategies

Smart campus technologies have huge opportunities to improve the learning environment and operational efficiencies throughout the whole institution, smoothing out many of the administrative functions. On the other hand, colleges and universities face many regulatory challenges in their way to adopting these new technologies to fullest use. These are challenges that majorly point towards cybersecurity and data security matters, environmental compliance issues; all these require due diligence to maintain compliance with legally prescribed requirements for protection of the stakeholders (Qin, 2023; Luckyardi et al., 2023).

The challenges of cybersecurity-related issues remain among the most immediate challenges in deploying smart campus technologies. It has resulted in a wide network ranging from sensors and cameras to any other kind of IoT devices, which has brought vulnerability for cyberattacks. Actually, colleges are going to be a major attraction in which there will be higher volumes of data related to personal, financial, and academic purposes that hackers love. Therefore, educational institutions should offer great security measures against intrusion into their networks, data breaches, and other malicious cyber activities: encryption protocols, periodic security audits, updating on the latest cybersecurity threats and solutions. Besides, colleges are presumed to consider a lot of legislation, such as the GDPR of the EU and the Family Educational Rights and Privacy Act of the United States, which put very strong measures of protection on students' and faculty members' data (Calza et al., 2022, Agriz et al., 2024).

Another challenge to the implementation of intelligent campus technology in colleges is data security. Because these intelligent systems create a host of data at higher learning institutions – student attendance, academic performance, behavioural analytics, and campus safety monitoring – data protection regarding such information becomes so critical. Data breaches and poor handling of personal information have serious reputational consequences, even legal ones. Also applicable in colleges and universities is an array of the general privacy laws and industry standards applicable to health care data under the Health Insurance Portability and Accountability Act, better known as HIPAA; likewise, that involving the information concerning the minors under COPPA (Children's Online Privacy Protection Act). The concept now calls for elaborate data governance frameworks that detail how to go about collecting, storing, accessing, and sharing such data in a way that only authorized persons would interact with the data, having kept all the privacy laws intact, including transparency to students, faculty, and staff (Bukhari et al., 2021; Huang, 2021).

Colleague campuses adopting the use of smart campus technologies face major regulatory challenges involving environmental compliance issues.

Many of the involved technologies set up smart campuses in relation to energy-efficient buildings, automation of lighting systems, and establishment of electric car charging stations. The majority are bound by various environmental policies at both the local and national levels. Most colleges have to implement set standards provided by the environmental protection agencies most of them relate to the regulation of energy consumption, emission, and electronic waste disposal (Rinaldi et al., 2018). For example, the introduction of energy-saving technologies requires the meeting of certain energy performance thresholds that are constrained by national or regional environmental policies. Besides that, colleges have to bear in mind the environmental impact of the gadgets they realize. This said, the obsolete devices should be disposed or recycled following related legislation of waste management, not to incur possible legal penalties and reduce environmental impacts linked to smart campus infrastructure (Trombin et al., 2021; Mascarenhas et al., 2021).

These are not mutually exclusive regulatory hurdles, and the colleges have to work their way through the overlap in deploying smart campus technologies. For example, deploying cybersecurity protocols might relate to demands for data security, and both have an impact on the college's environmental standards if, for example, certain technologies use too much energy or require methods of disposal that are out of compliance. It calls, therefore, for a holistic approach by the college in handling these regulatory challenges through teams that should bring on board experts in IT, compliance expertise, environmental science, and data management (Gupta et al., 2024; Chauhan et al., 2024).

Continuous monitoring and access by all institutions are recommended in light of regulatory changes and the emergence of new threats with regard to cybersecurity, data privacy, and environmental sustainability. Given that, cooperation can also be extended with various other players externally – for example, government agencies and any regulating authorities, technology providers in the private sector – who would support the institution in keeping pace with the relevant laws and best practices. In return, a proactive and integrated approach to the subject of regulatory compliance will empower colleges to answer such challenges and make full use of the potential smart campus technologies have in store for them while ensuring safety, privacy, and sustainability in campus environments (Liu and Liu, 2024).

Such is the balance that university innovation and research centres must achieve: leading-edge technological development in the face of a myriad set of diverse regulatory requirements. Increasingly, smart technologies deployed on campus – from AI to IoT and data-enabled infrastructure – offer new possibilities in improving the student experience, optimally managing resources, and campus safety. While not to disappoint in promising great ethical and legal issues testing universities in their way of meeting the innovation objectives and regulatory obligations, this really means the university will have a great challenge on how it should balance innovation with the ethical implications of the technologies they are going to deploy. Therefore, universities stand at the very forefront of knowledge and progress (Qin, 2023). Quite a number of universities identified themselves with the need to start testing those pioneering technologies, promising to bring edges of revolution within their living and learning processes. At the same time, most of these innovations are about collecting and processing serious amounts of personal data regarding their students, faculty, and employees. Therein lies a big question that touches on ethics: personal data protection and consent. That it means universities should have instituted data protection principles in the use of smart technology, transparency, fairness, and accountability. The informed consent entails it has to attain proper consent from all those individuals whose data it has collected, how it will make use of it, and opting out if they so decide (Kumari et al., 2023; Frias-Castillo et al., 2023; Chagnon-Lessard et al., 2021).

Besides this, it needs to balance the ever-growing demand for protection and promotion of privacy with the commitment to security. Most smart technologies require continuous data collection, ranging from student behaviour or campus attendance to even facial recognition for a raft of security reasons. While these may help achieve goals in operational efficiencies and campus security, they also raise very serious concerns about surveillance and other possible personal information misuses. To be sure, universities have had to apply strict laws in personal privacy, especially the GDPR in Europe and the Family Educational Rights and Privacy Act in the United States. These lay down very strict demands regarding what is to be done with the personal data of individuals, from the stages of its collection to its storage and dissemination. Ensuring that this legal framework has been followed secures protection against any legal liability and preserves the rights of a person (Calza et al., 2022; Kumari et al., 2023).

Lastly, there are quite a number of legal complications in terms of issues like ownership/IP of data emanating from such smart technologies. Nowadays, billions are invested by universities in research and development for new technologies. This has in turn brought up a dire need for some delineated policies regarding use and ownership with respect to intellectual property. Quite often, the legal landscape within which IP rights operate is complex insofar as collaboration among academic institutions, private companies, and other entities engaged in research is difficult to perceive. Of equal importance, universities should take into consideration how their partnership and agreements are framed in a manner to protect the interests of all parties while aligning innovations developed through smart technologies with values and mission of the institution (Gupta et al., 2024; Mascarenhas et al., 2021).

A further set of legal challenges that often has to be addressed in the implementation of smart technologies relates to the environmental and sustainability regulations. With increased awareness, universities have come under ever-increasing pressure to adopt and work towards sustainable practices – much of this extends to an institution's use of smart technologies. While energy-efficient buildings and automation systems represent ways to make a campus greener, they also need to align with and satisfy any local, national, or international standards set regarding the environment. It is also important that university policy will ensure the technologies adapted are not contributing to environmental degradation due to huge intake of energy resources, in fact to electronic waste improperly disposed. This will entail continued attention to ethical concerns and following of regulatory guidelines on sustainability (Frias-Castillo et al., 2023).

What this all sums up to is that operating against the background of such legal and ethical complications requires addressing innovation by starting from strategic approaches right from the time of entrance, so that entry would be accompanied by an integrated governance framework. In fact, for all intents and operational purposes, there is a need for an effective compliance management system that meets the legal and ethical expectations concerning responsible use of technology. That universities will need to be open with the way smart technologies are being deployed and made functional to help safeguard student, faculty, and staff rights by open discussion, possibly needing to involve even outside regulators. This calls for accountability and ethical responsibility, thus letting their technological advancement match up to goals of innovation and the regulatory standard that they are to adhere to (Chagnon-Lessard et al., 2021).

They further seek external expertise from legal advisors to ethicists, including the providers themselves, that the programs of their smart technologies do meet relevant laws and ethical requirements. Generally speaking, such engagement in development and management keeps universities prepared, not only for emerging regulations but also for associated ethics, rather than having their innovation agenda stopped by unexpected legal and ethical challenges. They would most likely stick to cautious planning, relentless monitoring, and dedication to accountable innovation; hence, the maintenance of regulatory

Table 10.2 The risk management and compliance initiatives in reducing possible legal and regulatory hazards

Risk management and compliance	*Description*	*Impact on legal and regulatory hazards*
Risk identification	Systematic process of pinpointing all kinds of potential risks in different areas of operation: data security, intellectual property, and environmental impact, among others.	Helps in considering and preventing legal challenges by pointing out areas that will lead to violations or breach of regulations.
Risk assessment	Assess the likelihood and potential consequence of identified risks on the operations and reputation of the university.	Gives a clear understanding of priorities of risks that need immediate consideration, reducing exposure to various legal and regulatory consequences.

(*Continued*)

Table 1.2 (Continued)

Risk management and compliance	*Description*	*Impact on legal and regulatory hazards*
Mitigation strategies	Devise ways of mitigating or eradicating the identified risks through more sophisticated security measures or the creation of policies.	Decreases the chances of incidents of legal violations and regulatory fines through early mitigation of each risk before it escalates.
Compliance with laws and regulations	Compliance with relevant legal frameworks that include but are not limited to data privacy laws, intellectual property rights, and environmental regulations.	Minimizes all legal risks by ensuring operations of the institution within boundaries set by regulatory requirements, avoiding penalties due to non-compliance.
Transparency and accountability	Accountability through clear policy, procedure, and practice communication and documentation.	It enhances confidence with regulatory bodies and stakeholders by reducing the risk of litigation against an organization due to non-compliance.
Training and awareness	Compliance requirement education for faculty, staff, and students; ethical standards.	It ensures that decisions are fully informed, avoiding unintended legal and regulatory violations.
Monitoring and auditing	Regular process and practice reviews to ensure ongoing compliance with laws and regulations.	It helps in early detection of compliance issues that can be dealt with before they turn into legal risks or fines.
Technology integration	Software and technology applications that allow for real-time monitoring, reporting, and managing of compliance activities.	It increases the efficiency and effectiveness of tracking compliance, reducing the risk of oversights leading to legal consequences.
Collaboration across departments	Coordinating legal, IT, compliance, and operations groups for a holistic approach in risk management.	Reduces legal risk by comprehensively addressing all potential hazards and ensures all aspects of compliance are covered.
External consultation and benchmarking	Engaging with lawyers, regulators, and peers regarding updates on changed regulations and the best ways to handle situations.	Helps to foresee regulatory changes and adjust internal policies in advance to keep them up to date with laws in constant evolution.

Source: Author's own work.

requirements and continued commitment by the university in technological advancement. Since leadership in education and technology lies with them, it shall be expected of them that the adoption of smart technologies elevates both academic and campus lives without breaching any legal and ethical consideration (Kumari et al., 2023; thrombin et al., 2021).

With large rosters of programs, departments, and vested interests, institutions of higher education are naturally exposed to a broad variety of risks ranging from cybersecurity and data privacy through intellectual property and environmental non-compliance. Risk management programs seek such risks – before they become major problems – in order that, on the institutional level, mitigants can be taken as far forward as possible (Table 10.2).

References

Abdulrasool, F., & Turnbull, S. (2020). The role of IT governance in enhancing the performance of smart universities. In *Advances in intelligent systems and computing, 1153 AISC* (pp. 708–720).

Ahriz, S., Gharbaoui, H., Benmoussa, N., Chahid, A., & Mansouri, K. (2024). Enhancing information technology governance in universities: A smart Chatbot system based on information technology infrastructure library. *Engineering, Technology and Applied Science Research*, *14*(6), 17876–17882.

Al-Jayyousi, O., & Aldhmour, F. M. (2023). Re-imagining the futures of university: Foresight for higher education using innovation labs learning space. In *Lecture notes in networks and systems, 495 LNNS*, pp. 318–327.

Bassano, C., Carotenuto, A., Ferretti, M., Pietronudo, M. C., & Coskun, H. E. (2019). Smart university for sustainable governance in smart local service systems. *Advances in Intelligent Systems and Computing*, *787*, 206–216.

Bukhari, E., Dabic, M., Shifrer, D., Daim, T., & Meissner, D. (2021). Entrepreneurial university: The relationship between smart specialization innovation strategies and university-region collaboration. *Technology in Society*, *65*, 101560.

Calza, F., Carayannis, E. G., Panetti, E., & Parmentola, A. (2022). The role of university in the smart specialization strategy: Exploring how university-industry interactions change in different technological domains. *IEEE Transactions on Engineering Management*, *69*(6), 2649–2657.

Chagnon-Lessard, N., Gosselin, L., Barnabe, S., Vandersickel, A., & Zhang, P. (2021). Smart campuses: Extensive review of the last decade of research and current challenges. *IEEE Access*, *9*, 124200–124234.

Chauhan, N., Karthikeyan, A., & Ravi Kumar, N. (2024). Transforming education: The intersection of education management, technology, smart universities, and the impact of Covid-19—A qualitative analysis, studies in systems. *Decision and Control*, *517*, 1103–1115/

Chen, Z., & Liu, Y. (2021). Research and construction of university data governance platform based on smart campus environment. *ACM International Conference Proceeding Series*, pp. 450–455.

CTM, Huawei and M.U.S.T. collaborate to launch cross-regional "5G Smart Campus" (2023).https://www.prnewswire.com/in/news-releases/ctm-huawei-and-must-collaborate-to-launch-cross-regional-5g-smart-campus-301996639.html (Accessed 5.02.2025).

Digital education action plan (2021-2027). (2020). https://education.ec.europa.eu/focus-topics/digital-education/action-plan (Accessed 5.02.2025).

Dodgson, M., & Staggs, J. (2012). Government policy, university strategy and the academic entrepreneur: The case of Queensland's smart state institutes. *Cambridge Journal of Economics*, *36*(3), 567–585.

Frías-Castillo, R., Flores-Amador, J., Romero-López, R., & Pérez-Hernández, P. (2023). *University-industry technology transfer in developing countries for smart cities, automation and innovation with computational techniques for futuristic smart, safe and sustainable manufacturing processes* (pp. 339–359).

Goals 4 ensure inclusive and equitable quality education and promote lifelong learning opportunities for all (2020). United Nations. https://sdgs.un.org/goals/goal4 (Accessed 5.02.2025).

Guenduez, A. A., Frischknecht, R., Frowein, S. C. J., & Schedler, K. (2024). Government-university collaboration on smart city and smart government projects: What are the success factors? *Cities*, *144*, 104648.

Gupta, S. K., & Rosak-Szyrocka, J. (2024). *Innovation in the university 4.0 system based on smart technologiesi* (pp. 1–230). New York: Chapman and Hall/CRC.

Huang, Z. (2021). An analysis of information security and protection strategies in colleges and universities in the environment of smart campuses. *Journal of Physics: Conference Series*, *1852*(4), 042050.

Jonek-Kowalska, I., & Wolniak, R. (2023). *Towards sustainability and a better quality of life?* London: Routledge.

Jonek-Kowalska, I., & Wolniak, R. (2025). *Exclusion in smart cities assessment and strategy for strengthening resilience*. London: Routledge.

Kangas, R., & Aarrevaara, T. (2020). Higher education institutions as knowledge brokers in smart specialization. *Sustainability*, *12*(7), 3044.

Kim, D., Chung, C.-S., Eom, S.-J., & Sung, W. J. (2024). Intelligent city: Opportunities, risks, and policy responses. In *ACM international conference proceeding series* (pp. 981–984).

Kumari, A., Rajput, S., Jain, A., & Kumar, S. (2023). Ai-based digital technologies in smart universities. In *Advancements in artificial intelligence, blockchain technology, and IoT in higher education: Mitigating the impact of COVID-19* (pp. 83–104).

Lamey, A., Abdelkader, H., Keshk, A., & Eletriby, S. (2023). A realistic and practical guide for creating intelligent integrated solutions in higher education using enterprise architecture. *Sustainability*, *15*(11), 8780.

Lee, H.-F., & Miozzo, M. (2024). Beyond complementarity and substitutability? Understanding relational governance and formal contracts in university-industry collaborations for innovation. *Technovation*, *138*, 103100.

Legal framework of EU data protection (2018). https://commission.europa.eu/law/law-topic/data-protection/legal-framework-eu-data-protection_en, (Accessed 5.02.2025).

Lian, J., Chen, J. (2024). Research and practice of data governance in universities based on data middleware. In *ACM international conference proceeding series* (pp. 238–242).

Liu, Y., & Liu, Z. (2024). Smart digital technology driven experience design of job seeking social skills training for university students in campus scenarios via business model innovation. In *Lecture notes in computer science (including subseries lecture notes in artificial intelligence and lecture notes in bioinformatics), 14716 LNCS* (pp. 117–127).

Liu, Z. (2020). Practice research on university data center construction from the perspective of data governance. In *Proceedings - 2020 2nd International conference on information technology and computer application, ITCA 2020* (pp. 489–492), 9422022.

Luckyardi, S., Hurriyati, R., Disman, D., & Dirgantari, P. D. (2023). Smart university image: Branding strategy in private universities. *Journal of Eastern European and Central Asian Research*, *10*(6), 929–939.

Ma, Y. (2024). The practical path of ruling colleges and universities according to law in the new era combined with big data analysis. *Applied Mathematics and Nonlinear Sciences*, *9*(1), 20231622.

Macao Institute of Smart City (2024). https://www.must.edu.mo/en/misc/intro (Accessed 5.02.2025).

Man, W., Zhu, Y., & Zhang, J. (2023). Analysis of data loss and disclosure in the process of big data governance in colleges and universities in the smart era. *Lecture Notes on Data Engineering and Communications Technologies*, *153*, 1453–1462.

Mario, J., Maja, C., & Tea, M. (2021). Relating smart governance as a university feature to students' university perceptions. *Journal of Information and Organizational Sciences*, *45*(1), 1–20.

Mascarenhas, C., Marques, C. S., Ferreira, J. J., & Galvão, A. R. (2021). The influence of research and innovation strategies for smart specialization (Ris3) on university-industry collaboration. *Journal of Open Innovation: Technology, Market, and Complexity*, *7*(1), 82.

Min-Allah, N., & Alarashed, S. (2020). Smart campus—A sketch. *Sustainable Cities and Society*, *59*, 102231, https://doi.org/10.1016/j.scs.2020.102231

Mircea, M., Stoica, M., & Ghilic-Micu, B. (2021). Investigating the impact of the internet of things in higher education environment. *IEEE Access*, 9, 33396–33409, 9359788.

Nguyen, T. T., & Nguyen-Dinh, C. H. (2024). A comparative analysis of development policies for smart universities in Vietnam. *International Journal of Innovative Computing, Information and Control*, *20*(3), 665–676.

Pham, T. V., Nguyen, A. T. T., Ngo, T. D., Le, D. H., Le, K. C. V., Nguyen, T. H. N., & Le, H. Q. (2020). Proposed smart university model as a sustainable living lab for university digital transformation. In *2020 5th International Conference on Green Technology and Sustainable Development (GTSD)*, Ho Chi Minh City, Vietnam, pp. 472–479.

Pinto, H. (2024). Universities and institutionalization of regional innovation policy in peripheral regions: Insights from the smart specialization in Portugal, *Regional Science Policy and Practice*, *16*(1), 12659.

Pirlone, F., Spadaro, I., & Candia, S. (2020). Smart policies for university students' sustainable mobility, *Territorio, (79)*, pp. 132–139.

Polin, K., Yigitcanlar, T., Limb, M., & Washington, T. (2023). The making of smart campus: A review and conceptual framework. *Buildings*, *13*(4), 891.

Qin, D. (2023). The demand and strategies of smart classroom construction - taking an international studies university as an example. In *ACM international conference proceeding series* (pp. 129–136).

Rinaldi, C., Cavicchi, A., Spigarelli, F., Lacchè, L., & Rubens, A. (2018). Universities and smart specialisation strategy: From third mission to sustainable development co-creation. *International Journal of Sustainability in Higher Education*, *19*(1), 67–84.

Salomaa, M., Charles, D., & Bosworth, G. (2023). Universities and innovation strategies in rural regions: The case of the Greater Lincolnshire Innovation Programme (UK). *Industry and Higher Education*, *37*(1), 67–79.

Santos, D. (2019). Smart cities and entrepreneurship: A new challenge for universities. In *Lecture notes in intelligent transportation and infrastructure, Part F1405* (pp. 118–131).

Smart campus (2024). https://sustainable.stanford.edu/smart-campus (Accessed 5.02.2025).

Snider, M. (2022). *Building a "smart" campus with care: Contradictions and communication in infrastructuring*. https://digital.lib.washington.edu/server/api/core/bitstreams/ecf7168e-e9f6-49fb-acd9-c933cf834a9a/content (Accessed 5.02.2025).

Soni, V., Sharma, A., Dhomase, K., Rajpurohit, P., & Shedge, R. (2023). UniDAO: University decentralised autonomous organization. In *2023 7th international conference on computing, communication, control and automation, ICCUBEA 2023*.

Spanoudakis, N. I., Krinakis, P., & Kolokotsa, D. (2024). A methodology for applying decision policies into smart buildings with the use of computational argumentation and iot technologies. In *EEITE 2024 - proceedings of 2024 5th international conference in electronic engineering, information technology and education*.

Stecuła, K., & Wolniak, R. (2022). Advantages and disadvantages of e-learning innovations during COVID-19 pandemic in higher education in Poland. *Journal of Open Innovation: Technology, Market, and Complexity*, *8*(3), 159.

Sustainable smart campus as a living lab (2024). https://ssc.hkust.edu.hk/, (Accessed 5.02.2025).

Telino, V., Massa, R., Mota, I., Gomes, A., & Moreira, F. (2020). A methodology for creating a macro action plan to improve it use and its governance in organizations. *Information*, 11(9), 1–27, 427.

The digital transformation of M.U.S.T.: Boosting innovation and building a smart campus (2024).https://e.huawei.com/en/case-studies/industries/education/2022/macao-university-science-technology (Accessed 5.02.2025).

Tomasi, S., Szávics, P., Aleffi, C., Cavicchi, A., & Hudec, O. (2024). Drivers and challenges of RIS3-related university engagement: Insights from five European regions. *Regional Science Policy and Practice*, *16*(1), 12567.

Tomás-Miquel, J.-V., Capó-Vicedo, J., Capó-Vicedo, J., & Mataix Domínguez, A. (2023). Beyond technology in the transition to smart campuses: The importance for policy-makers of the university community's perception. *Innovation: The European Journal of Social Science Research*, *36*(3), 361–380.

Trombin, M., Veglianti, E., Pinna, R., & Musso, M. (2021). Best practices and strategies in sustainability and smart mobility: Student engagement in Italian universities. *Lecture Notes in Information Systems and Organisation*, *36*, 47–63/

Tse, R., Mirri, S., Tang, S.-K., Pau, G., & Salomoni, P. (2021). GoodIT 2021, *Proceedings of the 2021 conference on information technology for social good* (pp. 309–312).

Wolniak, R. (2023). European Union Smart mobility—aspects connected with bike road system's extension and dis-semination. *Smart Cities*, *6*, 1009–1042.

Wolniak, R., & Stecuła, K. (2024). Evaluation of quality of innovative e-learning in higher education: An insight from Poland. *Applied System Innovation*, *7*(6), 109. https://doi.org/10.3390/asi7060109

Wolniak, R., Gajdzik, B., Grebski, M., Danel, R., & Grebski, W. W. (2024). Business models used in smart cities—theoretical approach with examples of smart cities. *Smart Cities 7*, 1626–1669.

Wolniak, R., & Jonek-Kowalska, I. (2022). The creative services sector in Polish cities. *Journal of Open Innovation: Technology, Market, and Complexity*, *8*, 17.

Zagrajek, K., Kłos, M., Rasolomampionona, D. D., Kołaczyński, P., & Suchecki, W. (2023). Investing in distributed generation technologies at Polish university campuses during the energy transition era. *Energies*, *16*(12), 4575.

11 The Future of Smart Universities in Smart Cities

11.1 Emerging Technologies Shaping Smart Universities

A significant element of the development of technology is the implementation of artificial intelligence (AI) and quantum computing in the curriculum in universities. Quantum computing is a completely new form of computation that can solve complex problems at extraordinary speed using fundamental physics principles. A bit is a unit of information, either 1 or 0 can be stored in that. On the other hand, quantum bits, or qubits, which have the ability to store zeros and ones, are the foundation of quantum computing. One fundamental characteristic of all quantum states is superposition, which is the ability of qubits to concurrently represent any combination of zero and one on classical computing, qubits are stored on physical hardware called chips (2023, 2025).

Quantum computing is one of the most important developments in higher education. Students can already explore how quantum ideas impact abstract maths and physics. This technology exploits the remarkable principles of quantum physics in order to perform calculations that are far beyond the scope of traditional computers. Together, quantum computers, which are able to solve some complex problems an almost inconceivable magnitude more quickly than it is possible for even the most powerful supercomputers to do, will supersede normal computers, which process binary data using transistors (Fernandes, 2024). They may ultimately allow for a better analysis of the information about students, allowing for adjustments to fit their needs and assist in assigning personalized attention to them.

Where currently available forms of adaptive learning models already support the construction of a personalized higher education program for each student, coupled with the real-time adaptation to their learning style and writing using content temporarily created to reflect recent publications, research, and extension, when computational power comes from quantum computers adaptive forms of learning could be supported. All research is going to be transformed by the integration of AI and quantum computing. Research could be done more comprehensively, and results could be revealed more timely (Fernandes, 2024).

DOI: 10.4324/9781003621829-12

By learning about AI and quantum computing, students are better prepared to not only use new technologies but also to help shape their future. This integrates and extends beyond skills for the workforce in supporting future generations as active contributors and participants in a world, and groups of people, are continuously redefining what is possible. The pace of advance in both fields predicts a day when AI and quantum computing will figure more prominently in problem-solving and technological advancement.

The quantum information science and technology (QIST) workforce is expected to grow 26% between 2016 and 2026. This level of growth is much higher than the average for all other professions, demonstrating the increasing demand for education and training in the sector. Table 11.1 presents case studies of universities that offer highly recognized courses in quantum computing.

The Rensselaer Polytechnic Institute (RPI), located in Troy, New York, USA, is a prestigious university that specializes in scientific research and is situated on the IBM computer campus. Employees and students can use

Table 11.1 Case studies of universities that offer highly recognized courses in quantum computing

University	*Description*
Massachusetts Institute of Technology (MIT) (USA)	Getting an education in technology at MIT ranks among the best in the world. All to address, essentially, at a hands-on basis, quantum algorithms, quantum hardware development and applications, all primarily geared towards a complete curriculum to prepare students for state-of-the-art research.
University of Cambridge (UK)	Through its internationally respected Department of Physics, the Top Universities for Quantum Computing of Cambridge contains one of Europe's best quantum computing programs. In collaboration with Quantum Motion, the university specializes in scalable careers in quantum computing technology and software-agnostic designs that create well-rounded graduates.
National University of Singapore (NUS) (Singapore)	In recent years, Singapore/NUS was the only Quantum Computing Programs offered in Asia. Its Centre for Quantum Technologies (CQT) is creating an integrative program covering physics, quantum computing education, computer science, and engineering for applications in the real world.
ETH Zurich, (Switzerland)	ETH has solid research bases and theoretical and experimental foundations on quantum mechanics and computation.
University of Waterloo (Canada)	The University of Waterloo is called Quantum Valley, which is the top school to learn quantum computing, and it's the number one in Canada as far as advanced infrastructure and quantum labs go worldwide.

Source: own study based on Samradni (2024).

quantum computers to work on problems in physics, chemistry, material science, and beyond (Afifi-Sabet, 2024). By 2030, the quantum computing market is expected to be valued at $125 billion with a CAGR of 33.2% over a forecasted period of 2022–2030. Organizations are now focusing on Quantum, as universities prepare the new generation of quantum scientists and engineers (Samradni, 2024).

Technologies like blockchain are being adopted by universities around the world to help manage significant quantities of research data. Blockchain is a cryptographic secure storage technology enabling tamper-proof data integrity to maximize the data output from large research projects (Singh, Rosak-Szyrocka, & Tamàndl, 2023). Blockchain technology can revolutionize the educational process especially from the perspective of database security, effectiveness and transparency (Lutfiani et al., 2021; Pu & Lam, 2023). It could change the way that educational accreditation and verification are done now. Its distributed digital ledger technology has transformational power to popularize the manner digital information is maintained and validated. A major use case is educational credentials. Blockchain increases accessibility for learners, institutions and employers. Education credentials, once tagged with the blockchain, become independent of location or institution, so they can be recognized anywhere and verified through learning institutions round the globe. Accordingly, educational credentials complement the vision of data democratization. Data democratization is the capacity to enable sharing of large datasets across institutions and researchers in all parts of the world. Blockchain democratizes access to large sets of data through distributed ledger technology, which allows for a wider breadth of participation in scientific research and education. Establishing blockchain in higher education is also expected to cut costs, making traditional, expensive centralized systems for verifying credentials and managing data obsolete. Streamlining these processes with blockchain can lead to more careful resource allocation, potentially lowering administrative costs and making education more accessible to students. University of Utah's Scientific Computing and Imaging Institute (SCI) is participating in the National Science Data Fabric (NSDF) and National Data Platform (NDP) initiatives, to facilitate global and equitable access to large scientific datasets — ranging to multiple petabytes of NASA climate data — bringing blockchain to forefront of academia. Using blockchain as an open platform for interactive processing and exploration of huge datasets without significant local resources, the university is breaking down past barriers to data access. Such effort expands the range of scientific inquiry, promotes an inclusive academic space, and demonstrates the power of blockchain to democratize access to information and create a culture of collaborative research (World Economic Forum, 2025).

When analysing the future of smart universities, it is worth paying attention to blockchain technology, which has enormous potential to revolutionize the education sector, especially in terms of data security and efficient academic record maintenance. Blockchain also enables a decentralized platform for educational transactions. With the help of this technology, the universities can maintain the security and efficiency of sharing sensitive student data using the cryptographic security and decentralization properties of blockchain technology. It will reduce the chances of security incidents and data breaches where educational institutions are targeted more in cyberattacks with the worse effect.

Academic institutions can also apply blockchain technology to create and authenticate micro-credentials as evidence of specific skills and competencies (Aulia & Yazid, 2021; Mourtzis, Angelopoulos, & Panopoulos, 2023; Shi, Shahriar, Lo, & Chi, 2022). This can forever change how educational data are shared securely and efficiently using blockchain technology. The direct example of this may be the development of a distributed ledger system in which administrators, teachers, and students can safely store and access data. Blockchain structure enables to share educational data fairly among several parties rather than relying on a single authority. Blockchain guarantees data integrity and accuracy. When the data is entered into the system, it cannot be modified or corrupted. This function guarantees that all stakeholders have access to the same and most recent information. Additionally, blockchain allows for a secure way of storing and sharing documents and transcripts.

An encrypted data is difficult to access without appropriate information (Chen, Xu, Lu, & Chen, 2018; Kosasi, Rahardja, Lutfiani, Harahap, & Sari, 2022). Additionally, blockchain could create a student's digital identity by storing academic records and achievements and providing access to diplomas, certificates, and other essential information. For this reason, it gives a secure, effective way of accessing and exchanging educational data between instructors and students (Alaa Abid Muslam, Mabrouk, & Zrigui, 2022; Alshahrani, Beloff, & White, 2020; Zhao et al., 2023). Blockchain technology offers a decentralized ledger offering a new form of transactions, applying to secure data storage, digital bidding, payments and authorization management in educational transactions. This could revolutionize the way educational institutions store data and conduct transactions. Blockchain generates an unchangeable and secure ledger, allowing payments between educational institutions, improving the systems transparency, lowering transaction costs and simplifying procedures. Additionally, limiting access solely to approved workers ensures the secure storage of sensitive information and student records by means of blockchain. It also enables digital signatures, document verification, and inter-institutional communication. Moreover, blockchain also reduces the risk of fraud and errors that leads to enhanced workflow efficiency. In addition, blockchain can manage authentication procedures, ensuring that students and other stakeholders can safely access their accounts and instructional materials. This reduces the risk of identity theft and unauthorized access to data. According to experts, blockchain technology will be crucial to the continued development of the metaverse. The metaverse is a virtual reality space in which users can interact with one another and the computer-generated environment in real time. Table 11.2 shows case studies of universities with emerging technologies.

Table 11.2 Case studies of universities with emerging technologies

University	*Technology*	*Description*	*Achievements*
Massachusetts Institute of Technology (MIT) (USA)	AI, quantum computing, blockchain	Improving research on campus, the better functioning of university administration, personalized learning	AI research–based tools development, Blockchain use cases innovation, More quantum computing research
University of California Berkeley (USA)	Quantum computing for research, blockchain for administration, AI for campus services	Efficient administration, improving learning experiences, AI-driven AI research collaboration	AI solutions integrated to support students, blockchain utilized for security of data, breakthroughs in quantum computing
University of Cambridge (UK)	AI-based research platforms, Blockchain for educational credentials, Immersive learning systems	Research has significantly improved, student participation has grown, and educational certifications may be swiftly verified.	Utilized AI for research optimization, blockchain for credentialing, immersive learning platforms
University of Toronto (Canada)	AI-driven data analytics, Blockchain for secure transactions, Immersive learning tools	Advanced data analytics, increasing student experience, campus operations optimization	Offering students attractive services with AI, more interesting educational offer with immersive learning, integrated blockchain
National University of Singapore (NUS) (Singapore)	Smart campus using AI, Blockchain for secure data, Quantum computing for innovation	Teaching innovation, research capacity, student engagement improvement	AI-based campus management systems have been introduced, quantum computers have been used for research purposes, and blockchain security has been improved

Source: own study.

11.2 Long-Term Impacts on Urban Development

Smart universities are leveraged to foster urban development in rich ways, but one of the most significant is through **economic development contribution**. These institutions forge circles of entrepreneurship and innovation that draw investment and create jobs. They promote start-ups and high-tech industries that are important for local and national economy with research and innovation hubs. Smart universities provide the skilled labour required to support growing fields according to evolving job market demands through specialized training and skills development programs. In addition, engaging with the industry, government sector and international organizations establishes a diverse group of stakeholders that encourages knowledge transfer and commercialization of research products, impacting the surrounding urban and geographical region, and helps boost its economy. Such world-class smart universities are often moving the erstwhile traditional industries to knowledge-based economies, which enable host cities to remain competitive in today's globalized world.

Universities are vital to realizing creative, educational, socioeconomic, structural, and cultural outcomes too (Langford, Hall, Josty, Matos, & Jacobson, 2006; Zhu, 2015). They play an essential role in the urban living system and use their collective intellect and student facilities to promote local urban and technical development (Anselin, Varga, & Acs, 1997). Additionally, universities work as developers of industrial, technical, and urban policies, drawing in investors and changing the growth trends of certain areas (Ischinger & Puukka, 2009). These connections have a role in the environmental and physical development of cities (Yalcintan & Thornley, 2007). Time-dependent urban growth dynamics have led to the establishment of certain colleges in distant locations in an effort to enhance structural and socioeconomic circumstances (Akengin & Kaykı, 2013). Moreover, the spatiotemporal dynamics of urban built-up are greatly influenced by universities situated in both urban and rural areas (Rahman et al., 2024). Universities are also developing into well-equipped locations for the start-up and growth of businesses, particularly high-tech ones. University-founded technical schools, technological valleys, and business incubators enable the creation of new technologies, and major corporations industrialize and apply scientific and technical advancements (Daneykin, Erznkyan, Ivanova, Trifonov, & Belchik, 2021).

While economic growth is a primary driver behind its emergence, smart universities also play a central role in **social justice** by fostering inclusivity, accessibility, and equity in the urban environment. These institutions offer outreach programs and scholarships that offer marginalized and underprivileged communities opportunities for empowering through education and skill development. This allows for the bridging of socioeconomic gaps and enables individuals from varying belief systems to not only partake in both work and economics but flourish through it. They support public service delivery (such as legal aid, healthcare, and community development) to local communities addressing social challenges and the urban population's well-being. Additionally, the rich cultural landscape created by these institutions promotes intercultural dialogue and mutual understanding, enhancing social cohesion and creating a stronger sense of belonging and community in smart cities. Cities that have welcomed smart universities have seen long-term benefits beyond the socioeconomic push. As an illustration, the connection between

Cambridge in the United Kingdom and the University of Cambridge has been great for that place, with a number of technological breakthroughs occurring there, all due to the backing of the university with the innovation ecosystem. Boston, Singapore, Eindhoven, and others have leveraged their universities for urban renewal, social inclusion, and sustainable planning. The success stories of these cities are a testament to how smart universities as active initiators can influence the trajectory of our cities to make them more resilient, inclusive, and economically vibrant.

A plan for civic engagement through public lectures, continuing education programs, service learning, and community service programs may also support the university's development of the community in which it operates, as well as the cities and regions that surround it. Collaboration among diverse organizations, including institutions of higher education (Savan & Sider, 2003), may yield stronger civic indicators like improved educational outcomes, decreased crime, and more affordable access to public transportation systems. As an example, the Sustainable Toronto project is a community-based research initiative that brings together the University of Toronto, York University, the City of Toronto, and regional environmental organizations. Community-based research is a collective term for a variety of research techniques that engage community members and outside researchers in an investigation to better understand local problems. Using three different partnership models, consultative, contractual, and collaborative, the project delivered several benefits to the community and the project including building capacity, monitoring assessment, sustainability promotion, and successfully submitted grant application for future projects.

In an assessment of the project, it was also discovered that the collaboration of community organizations, local government, and higher education formed strong ties among the partners, which aided in the project's successful completion. There are several instances of universities forming cooperative relationships with their urban neighbours. The University of Illinois at Chicago's Great Cities Institute cultivates relationships between institutional stakeholders inside and outside the campus (Baiocchi & Lerner, 2007; Gupta & Rosak-Szyrocka, 2024). According to research on major American and European cities, universities were crucial in advancing knowledge and science and in attracting knowledge workers (Power, Plöger, & Winkler, 2010). These communities were able to shift from an industrial foundation to a more profitable one, which enhanced the social welfare of their populace. Developing industry in tandem with research goals, cultivating industry collaborations, and creating and maintaining graduates who contribute to future advancements are all important anchoring roles played by higher education institutions (Ideopolis: Knowledge city-regions, 2006). The interactions between the many players in a city, such as corporations, government, and higher education, have an impact on how cities develop throughout the industrialized globe.

Apart from various economic and social contributions, smart universities are addressing the need for **promoting environmental sustainability** in urban development. Through these, they can act as a role model for their local community by embedding sustainable development within their campuses, for example, by deploying renewable energy sources, implementing waste management systems and erecting energy-efficient buildings. These universities are focused on leading-edge research on climate change, sustainable urban planning, and green technologies — and are producing evidence-based solutions for helping cities, churches, and institutions move in the direction of environmental stewardship. Through partnerships with local governments and businesses, they often implement policies and projects focused on carbon reductions, resource efficiency, and green transportation. In addition, they promote a sustainable model of urban living by, through educational programs and awareness campaigns, encouraging a culture of environmental responsibility within their students, faculty, and surrounding community.

Universities serve as hubs for urban social, scientific, and cultural life and act as catalysts for the economic growth of nations and regions. They use technology for the sustainable development of regions and help to create public spaces. Numerous nations, including China (*Development of green finance as a modern trend of the world economy (on the example of China)*, 2019; Khudyakova & Urumov, 2021), the United Kingdom, Austria, the Netherlands, France, and others (; *University campus-the basis for creating a modern educational landscape*, 2017), focus on the growth of university campuses and the impact that universities have on the prosperous development of cities. For instance, Guangzhou is a member of a regional cluster and includes ten institutions linked by a shared sociocultural infrastructure, transportation, and engineering systems (*University campus-the basis for creating a modern educational landscape*, 2017). The urban environment's growth was accelerated in 2013 with the opening of the Vienna University of Economics' new campus in Austria (Daneykin et al., 2021). The development of city university campuses includes a variety of activities, including the use of distinctive landscape features, public lectures, education, culture, recreation, libraries, and cafes, as well as an improvement in the quality of the urban environment. One prominent example of university-city is Tomsk in Siberia — the cultural and intellectual centre of Siberia. This city is distinguished by a high degree of creative entrepreneurship in its educators, researchers, and students, all knitting together academic, technical, cultural, scientific, and inventive strands.

Smart Universities are hotbeds of innovation, research, and education – considerably contributing to the development of a city in its technological advancement, environmental sustainability, and social inclusion (Ahriz, Gharbaoui, Benmoussa, Chahid, & Mansouri, 2024; Franco, Crisostomi, Leccese, Mugnani, & Suin, 2025; Wolniak et al., 2024). Interdisciplinary in approach and with state-of-the-art technologies to support it, smart universities house answers for perhaps the greatest challenges besetting modern urban metropolises: resource optimization, equality of opportunity, and inclusive growth (Shahzad, Khan, & Iqbal, 2024).

The real significant contribution, however, lies in the manner in which advanced technologies are integrated into the process of urban development, together with data-driven decision-making. Smart universities collaborate with governments, businesses, and NGOs in designing and implementing smart infrastructures that efficiently utilize resources related to energy, water, and transportation (Wardat & Akour, 2024). For instance, researchers at smart universities make models for the implementation of renewable sources of energy, perform advanced waste management, and practice sustainable building to reduce carbon footprints (Alzahrani, 2024; Yu, 2024). Such innovations provide environmental sustainability and also enhance the quality of life in cities by having cleaner, more efficient, and resilient cities. Smart universities thus act as think tanks to promote social inclusion and fight against inequalities (Chauhan, Karthikeyan, & Ravi Kumar, 2024; Lotfi, Rodić, Labus, & Bogdanović, 2024).

Smart universities also play a great role in the emancipation of marginalized groups through equity issues both in education and access to resources (Gupta and Rosak-Szyrocka, Staggs, Wright, & Jarvis, 2022). This takes the form of open access education, scholarships for students from economically weaker sections of society, and community-based programs to help address the gap in access to post-secondary education and skills training (Lorini & Chigona, 2020). It is in this way that they make sure that people of all different socio-economic backgrounds have the opportunity to participate in and contribute to the evolving knowledge economy. Equally smart universities foster inclusive and diverse cultures of innovation and entrepreneurship (Jonek-Kowalska & Wolniak, 2025). Hence, with diversified resources including funds, mentors, and spaces for collaboration, they are able to support students and researchers in converting their ideas into tangible solutions for answers to several challenges thrown by society. Such innovations are targeted at accessible and affordable technologies and services for reduction in inequality and growth of shared prosperity. Another aspect of their contribution is policymaking and public engagement (McLean, Maalsen, & Lake, 2022).

Smart universities act as an intermediary between policymakers, industry leaders, and the general public through their making evidence-based recommendations and fostering dialogues on issues of concern (Aljoufie & Tiwari, 2022; Shahzad & Ahmad Khan, 2024). Through the conduction of research on socio-economic impacts brought forth by urban development projects, such projects are made to be designed in a manner that benefits all sectors of society, especially the most vulnerable (Davydova, Turchenko, Spivak, & Dubrovskaya, 2021). Such institutions also offer public seminars, workshops, and educational campaigns that emphasize sustainable practice, active citizenship, and community engagement. Smart universities feel that global problems, such as climate change to social injustice, must be answered with a local accent (Dumnić & Popadić, 2022).

They thereby cooperate with local governments and other relevant levels of stakeholders in governance to ensure that urban planning and resource management strategies appropriately reflect the particular needs of the communities served. Such proximity of approach makes strategies more effective because it instils a sense of ownership and participation by the citizens (Artiaga, Banatao, Baud, Zhao, & Zhou, 2020). Their contributions go further than the field of academia and right into the realms of the entire societal framework towards which cities shall be equitable, resilient, and ecologically sustainable (Dumnic, Popadic, Milicevic, Guttler, & Feric, 2020; Eisapour, Jafarpur, & Farjah, 2021). Table 11.3 shows just how smart universities drive long-term benefits in urban

Table 11.3 The examples of long-term benefits from smart universities

City	*Smart university*	*Long-term benefits*
Boston, United States	Massachusetts Institute of Technology (MIT)	Innovations in urban planning, smart technology research, and sustainable transport have revived the local economy, making the start-up ecosystem at MIT a global tech hub.
Cambridge, United Kingdom	University of Cambridge	Extensive collaboration on research regarding clean energy and bioscience; contribution to sustainable urban planning and development of advanced educational access programs.
Singapore	National University of Singapore (NUS)	Leading in smart city innovation, including digital infrastructure and sustainable housing projects; partnering with government on green technologies and urban policy.
Stockholm, Sweden	KTH Royal Institute of Technology	Developing sustainable technologies and solutions for urban mobility, Stockholm came to the front line among the leaders of green innovation and energy-efficient city planning.
Seoul, South Korea	Korea Advanced Institute of Science and Technology (KAIST)	Improved ITS, Digital Governance: Efficiency in Urban Energy, AI at the Core of City Management.
Berlin, Germany	Technical University of Berlin	Joint projects with renewable energy sources, eco-housing, waste management; inculcation of research led by a university in the citywide drive to sustainability.
Barcelona, Spain	Universitat Politècnica de Catalunya (UPC)	Innovation in Intelligent Urban Mobility: bicycle-sharing systems, tourism practices supporting the resiliency of the city, methods for handling large-scale information flows across the entire city.
Tokyo, Japan	University of Tokyo	Pioneer research in disaster resiliency, renewable energy, and AI; impactful contribution to smart infrastructure.
San Francisco, USA	Stanford University	The outstanding contribution of innovation with significant enabling contribution towards the creation of an innovative Silicon Valley economy, development of advanced green building technology, and contribution to a technology-powered development of urbanism.
Sydney, Australia	University of Sydney	Research sustainable urban design, integrating renewable energy, building climate-resilient infrastructure, developing partnerships in the management of local and global environmental challenges.

Source: Author's own work.

settings as contributors to technological advancement, sustainable development, and socio-economic growth. These have been leveraged by the towns in making innovative academic institutions face up to urban challenges and work towards vibrant forward-thinking communities.

11.3 Strategic Planning for the Future of Smart Campuses

The development of Smart University will be underlain by the principle of long-term planning, coupled with flexibility, especially regarding how institutions will adapt to changing environments of technological innovation and urban transformation. Responsiveness of a university to new emerging technologies and shifting needs of society essentially delineates relevance and effectiveness in turning out future generations into an increasingly interlinked world (Shahzad et al., 2024, Berdnikova, Aksinina, Shchepotkina-Marinina, Lugovkina, & Kabanov, 2023a).

At the juncture of education, technology, and urban ecosystems, no doubt the smart university will need an operational rethink with structural relook at various levels. First and foremost, technologies – most of which involve aspects related to AI, the Internet of Things, and blockchain – provide unimaginable new opportunities in teaching and research but, all the same, are improving administrative processes in ways unimaginable some years ago (Alkassem, Draou, Alamri, & Alharbi, 2022; Wolniak & Stecuła, 2024a, Wolniak & Stecuła, 2024b). While doing so, they are also required to upgrade the infrastructure, faculties, and curriculum continuously in order to impart whatever skill sets are necessary for the students to be successful in an ever-changing professional world. Universities must, while balancing urgency in integrating such tools, establish a workable framework that could look towards future development (Kawamoto, Kukhareva, Weir, Brooke, & Del Fiol, 2021; Stecuła, & Wolniak, 2022b).

The urban contexts in which universities are inserted presently undergo deep changes, driven by globalization, environmental challenges, and changes in demography. Smart Universities will have to be designed so as to respond to these emerging needs from cities and regions as centres of knowledge and innovation (Zhou, 2021). This is not just about deploying technology to enable connectivity and collaboration; rather, it is about sustainability, inclusiveness, and urban mobility. For example, universities can also serve as living labs in the testing and prototyping of smart city technologies and policies by embedding students and faculty in the solving of complex real-world problems (Berdnikova, Sergeeva, Safronov, Smagina, & Ianitckii, 2020).

Flexibility will thus allow turning institutions outwards in the right direction towards unexpected challenges, such as going online at incredible speed due to disruption at a global level, and build a culture of innovation that would test and scale new ideas and methods at will. While this gives vision to such efforts, long-term planning assures the stability that must also be provided. Strategic foresight on investment aligns technology, infrastructure, and partnerships with broader institutional goals and societal needs (George & Wooden, 2023).

In fact, accordingly, it goes without saying that the long-term development and growth of the smart campus will be through strategic alliances – from a great number of kinds of stakeholders, including local authorities and business leaders (Luckyardi, Rahayu, Adiwibowo, & Hurriyati, 2023). In fact, such mechanisms of collaboration are part and parcel of the assurance of Smart Campus performance in placing it appropriately in relation to wider social, economic, and environmental imperatives. These alliances involve design considerations, effective channels of communication, and shared goals for mutual benefits. One of the most basic strategies that work towards successful smart campus life involves relationship building with local authorities. The local government does, of course, have a very core role in playing matters related to urban planning, structuring the zoning regulations, developing infrastructure, and resourcing – all issues directly touching on questions of functionality and scalability relating to smart campus projects (Berdnikova, Aksinina, Shchepotkina-Marinina, Lugovkina, & Kabanov, 2023b). Good communication with municipal leaders will help connect university visions with strategic priorities of the city. This might be the role of positioning at universities as an active contributor to regional innovation, economic development, and sustainable growth (Gudkova, Glukhova, Yakusheva, Burenkova, & Treshina, 2021.

The smart campus will contribute to developing the region in this case. In joint projects, such as creating projects for urban sustainable mobility or increasing the number of smart city infrastructure components, mutual confidence will be reached to ensure university contribution to common development. Of importance, equally not less, will be the close collaboration with business leaders and captains of industry (Berdnikova, Frolova, Pavlova, Zmievskii, & Igoshina, 2021). These would add a lot of value in terms of technological know-how, funding opportunities, and real insight into accelerating the implementation of smart campus solutions. This could be achieved if the universities offer themselves as active collaborators with corporations by offering a platform of research and development in which state-of-the-art technologies are tried out and fine-tuned under controlled conditions. On the business side, too, academic expertise and highly skilled talent will be rewarding and provide a place where new products and services can be tried out. The universities tap into resources, tools, and funding that would be difficult to access. Over time, these partnerships will evolve programs for the students to meet new job markets and ensure a pool of applicants in major industries (Jonek-Kowalska & Wolniak, 2023; Rico-Bautista, Medina-Cardenas, Areniz-Arevalo, Maestre-Gongora, & Guerrero, 2020).

The other side of alliance development encompasses collaboration with community organizations, non-profits, and other civic groups. These actors bring their set of visions that help the universities understand and address the social dimension in their smart campus initiatives. Spaces for collaboration with community stakeholders create inclusiveness, equity, and responsiveness in the campus developments towards the needs of the larger population. These may be on creating programs dealing with the digital divide, sustainability practices that will benefit the community living around it, or educative and cultural initiatives aimed at strengthening ties within the community (Shahzad et al., 2024. These will have to be forged consciously, with attention paid to the presence of shared aims and mutually reinforcing interests. The universities have to create narratives through which they articulate, with clarity, how the interests of the different stakeholders are served through intelligent campus initiatives. For example, there is a need for them to articulate how the Intelligent Campus builds on and contributes to regional economic growth, environmental sustainability, and social welfare. It develops ownership and, thus, a progressive and fruitful partnership through workshops, forums, and advisory councils on a regular basis. For us, successful alliance in the academia domain can be always done through openness and thereby confidence that those universities have evoked. Accountability – meaning regular reporting on the progress of projects, fiduciary arrangements, an assessment of impact – would reassure the stakeholders that these investments will bear fruits and, in fact, that they would be continuing to make those. Simultaneously, they should be willing and prepared to use feedback from stakeholders to change in strategy where needed to reassure that they are committed towards inclusiveness in decision-making and to shared decision-making altogether (Zhou, 2021).

In that way, universities forming strategic partnerships with local authorities, business leaders, and other stakeholders will be contributing to the creation of conditions that enable their smart campus to continue its growth in a sustainable manner. Actually, this is just the very foundation for innovative, resilient, and inclusive developments that will make campuses evolve together with their surroundings to meet the changing global demand. By sharing a collaborative ethos and commitment to improvement, Smart Campuses will be vibrant centres of learning, innovation, and community involvement, with their top priority being service to future generations (Wolniak & Stecuła, 2024b; Stecuła & Wolniak, 2022a). Table 11.4 presents in tabular form for reference the methods, discussed hereafter, on how Smart Universities can create and continue influential alliances while assuring their result in mutually beneficial and long-term sustainability.

Table 11.5 was measuring those institutions which could merge strategic plans and academic and urban development for gaining lessons from such implementation (Academic Strategic, 2024; Mc, 2024; Allen, 2024; Sweeney, 2022; Wilson, 2022; Niemtus, 2019).

Table 11.4 Tactics for smart universities to build long-term alliances with stakeholders

Stakeholder group	*Tactics for forming alliances*	*Expected outcomes*
Local authorities	Smart universities can collaborate with the local government at a formal level in various initiatives, assuring the establishment of forums or advisory councils at least on a regular communication basis. Matching university initiatives with municipalities' strategic goals, including those for urban sustainability and smart cities, brings a mutual growing strategy. Probably, one of the foundational points of collaboration is an improvement in public infrastructure: transport and connectivity. This will also contribute to the support of the university's municipalities in the solving of municipal challenges through research and innovation solutions to build a relationship in synergy.	Such partnerships build trust and collaboration, speed up regulatory approvals, develop joint infrastructure projects that are beneficial for both the university and the community, raise the reputation and integration of the university within the region, among others.
Business leaders	This means building relationships with technology firms to provide opportunities in research and development. In addition, the university should be building internship and recruitment pipelines for students into companies that seek the best talent in the country. It also helps them jointly develop innovative programs and labs according to emerging industries, thus creating opportunities to create leading-edge educational experiences. Finally, sponsorships and funding for business of smart campus projects and innovation initiatives can be sourced.	These alliances bring advanced technologies and expertise, financial support for campus development, better job placements for students, and effective talent pipelines for businesses. Jointly branded innovation initiatives raise the profile of the university as a hub of progress.
Community organizations	The institutions can develop a friendly association with the community organizations through establishing inclusive forums in which members of the locale participate, and mutual interests are elicited. Collaboration may be affected through programs which address social issues like digital divide and sustainability. The Universities can make available Campus resources such as libraries and cultural events for sharing with the Community in building relations with them. Active participation in events around the community solidifies the University position as an integral member of the region.	These partnerships will further reinforce positive community–campus relations, grow participation in campus initiatives, and promote equitable and inclusive campus policies, policies that strengthen university impact in society and connect deeper into the community.

(*Continued*)

Table 11.4 (Continued)

Stakeholder group	*Tactics for forming alliances*	*Expected outcomes*
Non-profits	For this, partnerships can be made in the fields of education, health, and environment. Here, universities could jointly prepare a grant proposal with non-profits where the goals from both the non-profit and university perspective are identical. In addition, institutions of higher learning have tremendous potential to increase their presence within the ranks of the marginalized communities simply by leveraging the access of non-profit organizations.	The partnerships with the non-profits further develop campus programs, increase access to funding opportunities, and raise the profile of the university in social responsibility matters. Again, such collaboration emphatically reaffirms the institutional commitment to a response to some pressing needs in society.
Government agencies	They may also be involved with government agencies by participating in joint research in various regional and national problems, with policy recommendations emanating from the academia. Apart from that, this advocates funding of higher education and policy support, thus cementing their relationship with government agencies. Other areas of collaboration may involve cybersecurity, renewable energy, public health, and other current concerns that align with both national priorities and university competencies.	Such an alliance strengthens the role of the university in public policy, provides access to the government funding of the smart campus projects, and raises its role in solving societal challenges. Besides, such partnerships with government agencies enhance the prestige of the university nationally and regionally.

Source: Author's own work.

Table 11.5 Case studies of institutions that have effectively aligned their strategic plans with academic and urban development goals

Institution	*Strategic focus*	*Key initiatives*	*Outcomes and lessons learned*
Arizona State University (ASU)	Integration of innovation and sustainability into both academic and urban planning frameworks.	Development of research centres addressing urban sustainability and renewable energy. Partnerships with local governments for smart city projects. Implementation of green infrastructure across campus.	ASU's initiatives have bolstered its reputation as a leader in sustainability and innovation, demonstrating the value of aligning academic goals with urban development. The university's strong community engagement model provides a replicable framework for other institutions.
National University of Singapore (NUS)	Focus on becoming a global leader in smart campus innovation while enhancing urban connectivity and liveability.	Adoption of Internet of Things (IoT) technologies to create a smart campus. Collaboration with the government on urban mobility research. Programs integrating environmental consciousness into academic curricula.	NUS exemplifies the potential of leveraging technology for campus operations and research while aligning with urban development. Its success underscores the importance of public–private partnerships and a clear focus on global challenges.
University College London (UCL)	Commitment to addressing societal challenges through interdisciplinary research and urban collaboration.	Creation of research hubs focusing on sustainable cities and global health. Partnerships with London's city government for policy development. Public outreach initiatives fostering urban inclusivity.	UCL's emphasis on interdisciplinary collaboration has enhanced its academic reputation and strengthened its urban impact. Its integration of research, policy, and community engagement offers a comprehensive model for strategic planning.

(*Continued*)

Table 11.5 (Continued)

Institution	*Strategic focus*	*Key initiatives*	*Outcomes and lessons learned*
University of Melbourne	Alignment of educational and research goals with the city's cultural, economic, and environmental priorities	Expansion of cultural and academic programs into urban spaces. Partnerships with local businesses for innovative research. Implementation of green building standards in campus facilities.	The university's deep connection to Melbourne's urban identity highlights the importance of embedding academic initiatives within local cultural and economic contexts. Its approach showcases the benefits of integrating sustainability into infrastructure development.
Delft University of Technology (TU Delft)	Advancement of technological innovation with a focus on sustainable urban environments.	Establishment of living labs for smart city technology testing. Collaboration with industry on renewable energy solutions. Development of education programs tied to urban innovation.	TU Delft's initiatives demonstrate the effectiveness of integrating research with practical applications. The living labs model serves as a practical approach to testing and scaling urban innovations in partnership with stakeholders.

Source: Author's own work.

References

Academic strategic plan archive for 2023–2024 planning. Arizona State University. (2024). https://provost.asu.edu/sites/default/files/2023-07/Academic%20Plan%20Archive_2023-2024_June%202023.pdf?utm_source=chatgpt.com, (Accessed 27.01.2025).

Afifi-Sabet, K. (2024). *New York college becomes 1st university with on-campus IBM quantum computer that is 'scientifically useful'*. https://www.livescience.com/technology/computing/rensselaer-polytechnic-institute-ibm-university-on-campus-quantum-computer-scientifically-useful (19 January 2025).

Ahriz, S., Gharbaoui, H., Benmoussa, N., Chahid, A., & Mansouri, K. (2024). Enhancing information technology governance in universities: A smart chatbot system based on information technology infrastructure library, *Engineering, Technology and Applied Science Research*, *14*(6), 17876–17882.

Akengin, H., & Kaykı, A. (2013). 'ŞEHİRLEŞME ÜNİVERSİTE İLİŞKİSİNE BİR ÖRNEK: GAZİMAĞUSA. *Marmara Coğrafya Dergisi*, (28), 501–525. https://dergipark.org.tr/en/pub/marucog/issue/475/3943

Alaa Abid Muslam, A. A., Mabrouk, M., & Zrigui, M. (2022). A review: Blockchain technology applications in the field of higher education. *Journal of Hunan University Natural Sciences*, *49*(10). http://www.jonuns.com/index.php/journal/article/view/1197

Aljoufie, M., & Tiwari, A. (2022). Citizen sensors for smart city planning and traffic management: crowdsourcing geospatial data through smartphones in Jeddah, Saudi Arabia. *GeoJournal*, *87*(4), 3149–3168.

Alkassem, A., Draou, A., Alamri, A., Alharbi, H. (2022). Design analysis of an optimal microgrid system for the integration of renewable energy sources at a university campus. *Sustainability*, *14*(7), 4175. https://doi.org/10.3390/su14074175

Allen, J. (2024). 14 smart cities in the US: Leading innovation to watch. https://www.sandtech.com/insight/smart-cities-in-the-us/, (Accessed 27.01.2025).

Alshahrani, M., Beloff, N., & White, M. (2020). Revolutionising higher education by adopting blockchain technology in the certification process. In *2020 International conference on innovation and intelligence for informatics, computing and technologies (3ICT)* (pp. 1–6). IEEE.

Alzahrani, A.M.M. (2024). Utilization of block chain technology and smart contracts in the education procedure of universities, *International Journal of Intelligent Systems and Applications in Engineering*, 12(12s), pp. 652–661.

Anselin, L., Varga, A., & Acs, Z. (1997). Local geographic spillovers between university research and high technology innovations. *Journal of Urban Economics*, *42*(3), 422–448.

Artiaga, E., Banatao, J. A., Baud, L., Zhao, C., & Zhou, Z. (2020). Space for urban planning: Team project conclusions from the space studies program. In *Proceedings of the international astronautical congress, IAC*, 2020 October.

Aulia, V., & Yazid, S. (2021). Review of blockchain application in education data management. In *2021 2nd international conference on smart computing and electronic enterprise (ICSCEE)* (pp. 95–101). IEEE.

Baiocchi, G & Lerner, J 2007, Could participatory budgeting work in the United States?. *The Good Society*, *16*(1), 8–13. https://muse.jhu.edu/pub/2/article/237823/summary

Berdnikova, L. F., Aksinina, O. S., Shchepotkina-Marinina, E. V., Lugovkina, O. A., & Kabanov, P. A. (2023a). Algorithm for strategic management of project activities in smart university, smart innovation, systems and technologies. In *355 SIST* (pp. 173–181).

Berdnikova, L. F., Aksinina, O. S., Shchepotkina-Marinina, E. V., Lugovkina, O. A., & Kabanov, P. A. (2023b). Algorithm for strategic management of project activities in smart university, smart innovation, systems and technologies. In *355 SIST* (pp. 173–181).

Berdnikova, L. F., Frolova, V. A., Pavlova, S.V., Zmievskii, D. V., & Igoshina, N. A. (2021). Strategic analysis of smart university resource potential for management objectives. *Smart Innovation, Systems and Technologies*, *240*, 351–361.

Berdnikova, L. F., Sergeeva, I. G., Safronov, S. A., Smagina, A. Y., & Ianitckii, A. I. (2020). Strategic management of smart university development. *Smart Innovation, Systems and Technologies*, *188*, 293–303.

Chauhan, N., Karthikeyan, A., & Ravi Kumar, N. (2024). Transforming education: The intersection of education management, technology, smart universities, and the impact of Covid-19—A qualitative analysis. *Studies in Systems, Decision and Control*, *517*, 1103–1115.

Chen, G., Xu, B., Lu, M., & Chen, N.-S. (2018). Exploring blockchain technology and its potential applications for education. *Smart Learning Environments*, *5*(1), 1–10. https://link.springer.com/article/10.1186/s40561-017-0050-x

Daneykin, Y., Erznkyan, B. A., Ivanova, O. P., Trifonov, V. A., & Belchik, T. A. (2021). Forecasting the contribution of universities to urban development. In *International scientific and practical conference "Russia 2020 - a new reality: Economy and society" (ISPCR 2020)* (pp. 59–62). https://www.atlantis-press.com/proceedings/ispcr-20/125953187

Davydova, T., Turchenko, A., Spivak, I., & Dubrovskaya, T. (2021). Customer engagement as the basis for technology decisions in a smart city. *E3S Web of Conferences, 263*, 04015.

Development of green finance as a modern trend of the world economy (on the example of China). (2019).

Dumnić, B., & Popadić, B. (2022). An innovative approach to a more sustainable and reliable energy society demonstrated by renewable energy sources for smart sustainable health centers. In *University education and other public buildings, proceedings of international conference on smart systems and technologies, SST 2022* (pp. 11–17).

Dumnic, B., Popadic, B., Milicevic, D., Guttler, Z., & Feric, D. (2020). Renewable Energy Sources for smart sustainable health centers, University Education and other public buildings. In *Proceedings of 2020 international conference on smart systems and technologies, SST 2020* (pp. 207–212), 9264132.

Eisapour, A. H., Jafarpur, K., & Farjah, E. (2021). Feasibility study of a smart hybrid renewable energy system to supply the electricity and heat demand of Eram Campus, Shiraz University; simulation, optimization, and sensitivity analysis. *Energy Conversion and Management, 248*, 114779.

Fernandes, N. (2024). Embracing the future with quantum computing in higher education. *Forbes*, 25 January. https://www.forbes.com/councils/forbesbusinesscouncil/2024/01/25/embracing-the-future-with-quantum-computing-in-higher-education/ (19 January 2025).

Franco, A., Crisostomi, E., Leccese, F., Mugnani, A., & Suin, S. (2025). Energy savings in university buildings: The potential role of smart monitoring and IoT technologies. *Sustainability, 17*(1), 111.

George, B., & Wooden, O. (2023). Managing the strategic transformation of higher education through artificial intelligence. *Administrative Sciences, 13*(9), 196.

Gudkova, S. A., Glukhova, L. V., Yakusheva, T. S., Burenkova, D. Y., & Treshina, I. V. (2021). Modern approach for strategic development of smart universities: digitalization and knowledge export. *Smart Innovation, Systems and Technologies, 240*, pp. 327–337.

Gupta, S. K., & Rosak-Szyrocka, J. (2024). *Innovation in the university 4.0 system based on smart technologies* (pp. 1–230). New York: Chapman and Hall/CRC.

Ideopolis: Knowledge city-regions 2006. https://scholar.google.com/citations?user=zb0rx90aaaaj&hl=en&oi=sra

Ischinger, B., & Puukka, J. (2009). Universities for cities and regions: Lessons from the OECD reviews. *Change: The Magazine of Higher Learning*, 41(3), 8–13. https://www.tandfonline.com/doi/pdf/10.3.00/chng.41.3.08-13

Jonek-Kowalska, I., & Wolniak, R. (2023). *Towards sustainability and a better quality of life?* London: Routledge.

Jonek-Kowalska, I.; Wolniak, R. (2025). *Exclusion in smart cities assessment and strategy for strengthening resilience*. London: Routledge.

Kawamoto, K., Kukhareva, P.V., Weir, C., Brooke, B.S., & Del Fiol, G. (2021). Establishing a multidisciplinary initiative for interoperable electronic health record innovations at an academic medical center. *JAMIA Open, 4*(3), ooab041.

Khudyakova, L., & Urumov, T. (2021). "Green" finance in BRICS countries. *Mirovaia ekonomika i mezhdunarodnye otnosheniia, 65*(9), 79–87. https://ras.jes.su/meimo/s013122270016962-9-1-en

Kosasi, S., Rahardja, U., Lutfiani, N., Harahap, E. P., & Sari, S. N. (2022). Blockchain technology - emerging research themes opportunities in higher education. In *2022 International conference on science and technology (ICOSTECH)* (pp. 1–8). IEEE.

Langford, C. H., Hall, J., Josty, P., Matos, S., & Jacobson, A. (2006). Indicators and outcomes of Canadian university research: Proxies becoming goals?. *Research Policy, 35*(10), 1586–1598.

Lorini, M. R., & Chigona, W. (2020). ICT for development, empowerment, or emancipation? A reflection on terminology based on case studies in underserved communities of cape town. In *Proceedings of the 13th IADIS international conference ICT, society and human beings 2020, ICT 2020 and proceedings of the 6th IADIS international conference connected smart cities 2020, CSC 2020 and proceedings of the 17th IADIS international conference web based communities and social media 2020, WBC 2020 - Part of the 14th multi conference on computer science and information systems, MCCSIS 2020* (pp. 65–72).

Lotfi, F., Rodić, B., Labus, A., & Bogdanović, Z. (2024). Smart healthcare: Developing a pattern to predict the stress and anxiety among university students using machine learning technology. *Journal of Universal Computer Science*, *30*(10), 1316–1342.

Luckyardi, S., Rahayu, A., Adiwibowo, L., & Hurriyati, R. (2023). Information technology in evolutionary strategic management: Decision support system in smart university, *Journal of Engineering Science and Technology*, *18*(3), pp. 1561–1569.

Lutfiani, N., Aini, Q., Rahardja, U., Wijayanti, L., Nabila, E. A., & Ali, M. I. (2021). 'Transformation of blockchain and opportunities for education 4.0. *International Journal of Education and Learning*, 3(3), 222–231. https://pubs2.ascee.org/index.php/ijele/article/view/283

Mc Dannald, E. (2024). How smart cities can impact campus life. https://facilityexecutive.com/how-smart-cities-can-impact-college-campus-life/ (Accessed 27.01.2025).

McLean, J., Maalsen, S., Lake, L. (2022). Digital (un)sustainability at an urban university in Sydney, Australia. *Cities*, *127*, 103746.

Mourtzis, D., Angelopoulos, J., & Panopoulos, N. (2023). 'Blockchain integration in the era of industrial metaverse. *Applied Sciences*, 13(3), 1353. https://www.mdpi.com/2076-3417/13/3/1353

Niemtus, Z. (2019). *Are university campuses turning into mini smart cities?* https://www.theguardian.com/education/2019/feb/22/are-university-campuses-turning-into-mini-smart-cities (Accessed 27.01.2025).

Power, A., Plöger, J., & Winkler, A. (2010). *Phoenix cities. The fall and rise of great industrial cities*. Policy Press.

Pu, S., & Lam, J. S. L. (2023). 'The benefits of blockchain for digital certificates: A multiple case study analysis. *Technology in Society*, *72*, 102176. https://www.sciencedirect.com/science/article/pii/s0160791x22003177

Rahman, M. N., Saleheen, M. M., Ahmad, B., El Fadili, H., Sharifuzzaman, S. A. S. M., Sohel, M. S., Jahan, S. H., Sarker, M. F. H., Islam, A. R. M. T., & Azim, S. A. (2024). Transforming landscapes: Decoding the impact of universities on urbanization using advanced modeling and perception analysis. *PLOS ONE*, *19*(10), e0302362.

Rico-Bautista, D., Medina-Cardenas, Y., Areniz-Arevalo, Y., Maestre-Gongora, G., Guerrero, C. D. (2020). Smart university: Big data adoption model | universidad inteligente: Modelo de adopciÃn de big data, applications in software engineering. In *Proceedings of the 9th international conference on software process improvement, CIMPS 2020* (pp. 52–60). 9390151.

Samradni (2024). Innovation meets education: top universities offering quantum computing programs. *Analytics Insight*, 24 December. https://www.analyticsinsight.net/tech-news/innovation-meets-education-top-universities-offering-quantum-computing-programs (19 January 2025).

Savan, B., & Sider, D. 2003, 'Contrasting approaches to community-based research and a case study of community sustainability in Toronto, Canada. *Local Environment*, *8*(3), 303–316.

Shahzad, K., & Ahmad Khan, S. (2024). Effects of E-learning technologies on librarians' sustainable competence development and smart library services: empirical evidence from 244 universities' librarians, *Libri*, *74*(4), 393–407.

Shahzad, K., Khan, S. A., & Iqbal, A. (2024). Factors influencing the adoption of Internet of Things (IoT) in university libraries: A systematic literature review (SLR), *Electronic Library*, *42*(2), 255–287.

Shi, Y., Shahriar, H., Lo, D., & Chi, H. (2022). Enhancing blockchain technology education with innovative active learning. In *2022 IEEE 2nd international conference on advanced learning technologies on education & research (ICALTER)* (pp. 1–4). IEEE.

Singh, S., Rosak-Szyrocka, J., & Tamàndl, L. (2023). Development, service-oriented architecture, and security of blockchain technology for Industry 4.0 IoT application. *HighTech and Innovation Journal*, *4*(1), 134–156. https://www.hightechjournal.org/index.php/hij/article/view/410

Staggs, J., Wright, A.L., & Jarvis, L. (2022). Institutional change, entrepreneuring and place: Building a smart state. *Organization Studies*, *43*(2), 269–288.

Stecuła, K., & Wolniak, R. (2022a). Advantages and disadvantages of e-learning innovations during COVID-19 pandemic in higher education in Poland. *Journal of Open Innovation: Technology, Market, and Complexity*, *8*(3), 159.

Stecuła, K., & Wolniak, R. (2022b). Influence of COVID-19 pandemic on dissemination of innovative e-learning tools in higher education in Poland. *Journal of Open Innovation: Technology, Market, and Complexity*, *8*(2), 89.

Sweeney, E. (2022). *City-university partnerships are a win-win. Here's how they can best work together to fight climate change and adopt new tech.* https://www.businessinsider.com/how-cities-can-partner-universities-advance-innovation-efforts-2022-7?IR=T (Accessed 27.01.2025).

The influence of quantum computing on future curriculum developments – UCAT Utah College of Applied Technology (2025). https://www.ucats.org/the-influence-of-quantum-computing-on-future-curriculum-developments.html (19 January 2025).

University campus-the basis for creating a modern educational landscape (2017).

Wardat, S., & Akour, M. (2024). Adoption of smart watches as wearable technology in TESOL education among university students. *Smart Learning Environments*, *11*(1), 35.

What is quantum computing? (2023). *McKinsey & Company*, 01 May. https://www.mckinsey.com/featured-insights/mckinsey-explainers/what-is-quantum-computing [19 January 2025].

Wilson, S. (2022). *How smart urban areas can benefit everyone*. https://www.weforum.org/stories/2022/07/smart-cities-urban-areas-that-benefit-everyone/ (Accessed 27.01.2025).

Wolniak, R., Gajdzik, B., Grebski, M., Danel, R., & Grebski, W. W. (2024). Business models used in smart cities—theoretical approach with examples of smart cities. *Smart Cities*, *7*, 1626–1669.

Wolniak, R., & Stecuła, K. (2024a). Evaluation of quality of innovative e-learning in higher education: An insight from Poland. *Applied System Innovation*, *7*(6), 109. https://doi.org/10.3390/asi7060109

Wolniak, R., & Stecuła, K. (2024b). Artificial intelligence in smart cities—applications, barriers and future directions: A review. *Smart Cities*, *7*, 1346–1389.

World Economic Forum (2025). *How universities can use blockchain to transform research*. https://www.weforum.org/stories/2024/03/higher-education-universities-blockchain-transform-research/ [19 January 2025].

Yalcintan, M.C., & Thornley, A. (2007). Globalisation, higher education, and urban growth coalitions: Turkey's Foundation universities and the case of Koç University in Istanbul. *Environment and Planning C: Government and Policy*, *25*, (6), 822–843.

Yu, X. (2024). Research on smart contract application system integrating big data platform and blockchain technology in universities. In *2024 IEEE 2nd international conference on sensors, electronics and computer engineering, ICSECE 2024* (pp. 1198–1202).

Zhao, M., Liu, W., Saif, A.N.M., Wang, B., Rupa, R. A., Islam, K. M. A., Rahman, S. M. M., Hafiz, N., Mostafa, R., & Rahman, M. A. (2023). Blockchain in online learning: A systematic review and bibliographic visualization. *Sustainability*, *15*(2), 1470. https://www.mdpi.com/2071-1050/15/2/1470

Zhou, Z. (2021). The development strategy research of higher education management from the perspective of "Internet+". *Advances in Intelligent Systems and Computing*, *1303*, 444–450.

Zhu, C. (2015). 'Organisational culture and technology-enhanced innovation in higher education. *Technology, Pedagogy and Education*, 24(1), 65–79.

Index

Pages in **bold** refer to tables.

FSC
www.fsc.org
MIX
Papier | Fördert
gute Waldnutzung
FSC® C083411

Kolibri 360 GmbH
Ferdinand-Jühlke-Straße 7
99095 Erfurt, Deutschland
produktsicherheit@kolibri360.de